The Prospects for Increasing the Reuse of Digital Training Content

Michael G. Shanley, Matthew W. Lewis, Susan G. Straus,
Jeff Rothenberg, Lindsay Daugherty

Prepared for the Office of the Secretary of Defense
Approved for public release; distribution unlimited

RAND NATIONAL DEFENSE RESEARCH INSTITUTE

The research described in this report was prepared for the Office of the Secretary of Defense (OSD). The research was conducted in the RAND National Defense Research Institute, a federally funded research and development center sponsored by the OSD, the Joint Staff, the Unified Combatant Commands, the Department of the Navy, the Marine Corps, the defense agencies, and the defense Intelligence Community under Contract W74V8H-06-C-0002.

Library of Congress Cataloging-in-Publication Data

The prospects for increasing the reuse of digital training content / Michael G. Shanley ... [et al.].
 p. cm.
 Includes bibliographical references.
 ISBN 978-0-8330-4661-1 (pbk. : alk. paper)
 1. Distance education—Computer-assisted instruction. 2. Internet in education. 3. Instructional systems—Design. 4. Military education—United States—Computer-assisted instruction. I. Shanley, Michael G., 1947–

LC5803.C65P76 2009
371.3'58—dc22

2009006823

The RAND Corporation is a nonprofit research organization providing objective analysis and effective solutions that address the challenges facing the public and private sectors around the world. RAND's publications do not necessarily reflect the opinions of its research clients and sponsors.
RAND® is a registered trademark.

© Copyright 2009 RAND Corporation

Permission is given to duplicate this document for personal use only, as long as it is unaltered and complete. Copies may not be duplicated for commercial purposes. Unauthorized posting of RAND documents to a non-RAND Web site is prohibited. RAND documents are protected under copyright law. For information on reprint and linking permissions, please visit the RAND permissions page (http://www.rand.org/publications/permissions.html).

Published 2009 by the RAND Corporation
1776 Main Street, P.O. Box 2138, Santa Monica, CA 90407-2138
1200 South Hayes Street, Arlington, VA 22202-5050
4570 Fifth Avenue, Suite 600, Pittsburgh, PA 15213-2665
RAND URL: http://www.rand.org/
To order RAND documents or to obtain additional information, contact
Distribution Services: Telephone: (310) 451-7002;
Fax: (310) 451-6915; Email: order@rand.org

Preface

The Department of Defense (DoD) is interested in expanding the use of distributed learning (DL) for military training and in understanding how DL development might be encouraged through large-scale reuse of digital content. The RAND Corporation was asked to examine how the Advanced Distributed Learning (ADL) Initiative and DoD more broadly might encourage both reuse and the development of a learning object economy. The study focused on the supply side of the reuse market, especially how incentives (both economic and non-economic) and other enablers might be used to encourage training development organizations to develop reusable learning objects.

Four key questions guided the research:

- To what extent are training development organizations currently engaged in reuse at this stage of technological development?
- To what extent do organizations find reuse a worthwhile investment?
- To what extent do disincentives to wider sharing of learning objects impede reuse?
- To what extent do organizations know how to implement a reuse strategy?

This monograph summarizes the findings of the research. It should be of particular interest to those involved in training, training standards, distributed learning, training transformation, or the reuse of digital training content.

This research was sponsored by the ADL Initiative within DoD and the U.S. Joint Forces Command and was conducted within the Forces and Resources Policy Center of the RAND National Defense Research Institute, a federally funded research and development center sponsored by the Office of the Secretary of Defense, the Joint Staff, the Unified Combatant Commands, the Department of the Navy, the Marine Corps, the defense agencies, and the defense Intelligence Community.

For more information on RAND's Forces and Resources Policy Center, contact the Director, James Hosek. He can be reached by email at jrh@rand.org; by phone at 310-393-0411, extension 7183; or by mail at the RAND Corporation, 1776 Main Street, Santa Monica, California 90407-2138. More information about RAND is available at www.rand.org.

Contents

Preface iii
Figures ix
Tables xi
Summary xiii
Acknowledgments xxv
Abbreviations xxvii

CHAPTER ONE
Introduction 1
Reuse of Content Is a Strategy for Reducing the Development Costs
    of Distributed Learning 1
Research Questions Examined 2
Preview of Key Findings and Recommendations 3
Focus of Project: Supply Side and Market Enablers 4
Focus of Research Approach: Early Adopters of Reuse for e-Learning
    and Reuse in Other Contexts 6
Organization of This Monograph 8

CHAPTER TWO
**The Prevalence of Reuse and the Role of Standards and
    Technologies** 11
Organizations Pursue Reuse Using a Number of Approaches ... 11
RLO-Based Reuse Is Less Prevalent Than Other Approaches ... 15
Relatively New Technical Standards and Technologies for Reuse
    Partially Explain the Relative Scarcity of the RLO Approach ... 17
Concept Reuse Avoids the Technical Problems of RLO-Based Reuse ... 20

v

CHAPTER THREE
Economic Incentives 23
Few Training Organizations Had More Than Modest Returns from Pursuing Reuse 23
The Decision to Bypass an RLO-Based Reuse Approach Makes Sense for Many Training Development Organizations 26
Some Digital Markets/Repositories Have Succeeded, Others Have Failed 32
History of Attempts to Reuse Software Provides Insight on Training Content Reuse 37
Experience with Software Reuse Suggests That Success Will Occur Only in Selected Cases and That They Will Be Difficult to Identify Beforehand 39
History of Reuse in Materiel Design Provides Insight on Training Content Reuse 44
Recommendation: Make ROI for Reuse an ADL Focus Area 48

CHAPTER FOUR
Disincentives to Sharing 53
Various Stakeholders Participate in a Strategy of e-Learning Reuse 53
Training Development Organizations Considered Disincentives to Be Secondary Obstacles to a Successful Reuse Strategy 54
Disincentives May Become a Bigger Issue If Reuse in e-Learning Becomes More Prevalent 57
Incentive Mechanisms for Reuse Might Be Created by a Variety of Strategies 60
DoD Mandate to Reuse May Require Additional Incentives to Be Effective 62
Recommendation: Stimulate Additional Incentive Mechanisms for Participation in Reuse Strategies 63

CHAPTER FIVE
Implementation Issues 65
Implementing a Reuse Strategy for e-Learning Requires Significant Change Management 65

TD Organizations Identified Many Obstacles Relating to
    Implementation of a Reuse Strategy 68
What TD Organizations Saw as the Greatest Obstacle Depended
    on the Current Status of Their Reuse Efforts 72
A Supportive Environment Is Needed to Implement an Effective
    Reuse Strategy 74
Widespread Success with Reuse Requires Effective Collaboration
    Across Organizations 77
Recommendation: Provide Additional Support for Processes That
    Implement Reuse Strategies 79

CHAPTER SIX
Overall Recommendations... 83

APPENDIXES
A. **Case Study Results** 87
B. **Interview Protocol and Questions** 95

References 107

Figures

1.1.	Elements of an Exchange System for Reusable Learning Objects	5
2.1.	Extent to Which Different Approaches to Reuse Have Been Employed	15
2.2.	Place of RLO-Based Reuse in the Technology Adoption Life Cycle	19
3.1.	The Extent to Which Organizations Saved e-Learning Development Resources by Employing a Reuse Strategy	24
3.2.	Example: Distribution of Number of Graduates for High-Priority e-Learning Courses Within a Large Defense Organization	29
3.3.	Example: Orders from DAVIS/DITIS Digital Repositories Suggest That Only a Few Repository Items Get Substantial Reuse	31
4.1.	Stakeholders Associated with TD Organizations and Their Potential Relationship	55
5.1.	Implementation Obstacles Reported by TD Organizations	68
5.2.	Distribution of What Was Seen as the Greatest Obstacle to Reuse, by Status of Current Reuse Strategy	73

Tables

1.1. Organizations Participating in Formal Telephone Interviews 9
3.1. Successes and Failures of a Range of Digital Markets/Repositories 33

Summary

Distributed learning (DL) offers the promise of self-paced learning and training at any time and in any place, as well as new technologies for developing and delivering content and tracking student performance. Although demand for DL is increasing, DL still represents a small percentage of all learning and training, in part because of the high cost of developing and maintaining electronic-learning (e-Learning) materials. Development costs for DL might be reduced if digital content could be reused on a large scale—i.e., if existing digital content could be used to produce new content or applied to a new context or setting. One option for encouraging widespread reuse is to create and link learning object repositories—i.e., searchable databases in which digital content is stored in the form of learning objects and accessed by others to create new course content.

In 2006, RAND was asked to examine how the Advanced Distributed Learning (ADL) Initiative and the Department of Defense (DoD) more broadly might encourage reuse through the use of learning object repositories and the eventual emergence of a learning object economy. The study's primary focus was on the extent to which incentives and other enablers currently are and might be used to encourage training development (TD) organizations to develop a reuse mechanism (especially the supply side of it) supported by repositories. Four key questions guided the research:

1. To what extent are TD organizations currently engaged in reuse at this stage of technological development?

2. To what extent do organizations find reuse a worthwhile investment?
3. To what extent do disincentives to wider sharing of learning objects impede reuse?
4. To what extent do organizations know how to implement a reuse strategy?

To answer these questions, we conducted structured telephone interviews in late 2006 and early 2007 with individuals within a wide range of large TD organizations in both DoD and foreign defense organizations, as well as in other U.S. government organizations, the commercial sector, and academia. We also conducted site visits and more-extensive interviews at two of these organizations. In addition, we reviewed studies on incentive issues in the knowledge management literature and on reuse efforts in the domains of software and materiel development. Additionally, we interviewed experts in various aspects of reuse strategies (e.g., experts in digital rights management).

Key results of the study follow.

We Identified Five Types of Reuse in Training Development Organizations

Our initial research found that TD organizations used three primary strategies in pursuing reuse:

1. The *top-down (coordination-driven) approach*. The TD organization collaborates with other TD organizations on course design or otherwise coordinates so that e-Learning courses can reach wider audiences.
2. The *reusable learning object (RLO) approach*. The TD organization designs and reuses digital content as independent objects, complete with learning objective(s), interaction, and assessment.
3. The *bottom-up (asset-driven) approach*. The TD organization reuses digital assets (e.g., images, sound, video) directly in learning.

Our interviews revealed two additional strategies:

1. *Concept reuse.* The TD organization reuses pedagogical approaches, including instructional methods, task decomposition approaches, and assessment methods. This reuse strategy is similar to a researcher's use of papers on related research as models for the design of a new study.
2. *Structural reuse.* The TD organization adopts some type of development structure, be it as simple as a template or style sheet or as complex as a content management environment (e.g., one of the commercially available learning content management systems that allow users to create and reuse digital learning assets and content within a common authoring environment).

Reuse Is Occurring, But Reuse Based on the Reusable Learning Object Approach Is Relatively Rare, and Technical Challenges Will Take Time to Overcome

We found that, at the time of our interviews, the RLO approach to reuse was less prevalent than the top-down or bottom-up approaches. Roughly 20 percent of the organizations interviewed reported successful reuse with the RLO approach. This number seems particularly low, given that we sought out organizations having the greatest experience with reuse. In contrast, 70 percent of the TD organizations reported using the bottom-up (asset-driven) approach, and 85 percent used some form of the top-down (coordination-driven) approach. Although some reuse approaches involved sophisticated collaboration, the most prevalent form overall was simple redeployment of entire courses.

One reason for low use of the RLO approach is that although technical standards for sharing content are well established, adoption of these standards is not yet complete, and improvements in interoperability are still needed. Moreover, authoring technologies and content management systems (CMSs) are evolving but are not yet to the point of being cost-effective for a wide range of potential users. In general, we concluded that technologies that support reuse are in the earliest stage of the technology-adoption life cycle, and progress toward widespread adoption is likely to be relatively slow.

Given the technical challenges that potential reusers presently face, we think it important that the concept reuse approach not be overlooked. Concept reuse needs to be acknowledged so that it can be measured and documented as part of the early success with reuse and can be supported in the design of large-scale repositories. In particular, since the success of concept reuse depends on being able to quickly locate content and explore it for possible emulation, there is a need for a capability that quickly searches for and accesses content or content summaries for inspection.

Significant Returns from Reuse Are the Exception, and Successes Will Remain Difficult to Predict

Our interviews suggest that few TD organizations view their return on investment (ROI) from reuse as anything more than modest, even after several years of pursuing an RLO-based reuse approach. Only 25 percent of the organizations interviewed estimated a positive ROI in line with their expectations, and these organizations typically used either the top-down or the bottom-up approach to reuse, or both. The majority of organizations estimated that they had achieved lower than expected returns, and 35 percent reported no savings at all or a net loss.

Two organizations reported large savings from efforts to restructure their development environment—i.e., from structural reuse. These results mirror case study findings from the commercial sector that show large savings from adopting technologies that automate the reuse of content in multiple delivery formats (e.g., online courses, job aids, instructor guides, lesson plans).

For many organizations, the decision to bypass an RLO-based reuse strategy appears to make sense economically. Implementation of an RLO-based reuse initiative requires significant up-front investment and organizational change, and any returns are at best years away and by no means guaranteed. For example, the demand for existing content has proven in many contexts to be too small to justify the investment in reuse. Moreover, service to immediate customers can sometimes be compromised by the redirection of efforts and resources toward reuse outside those customers' interests. Finally, other approaches for reducing development costs, including rapid authoring methods and internal process improvements, can sometimes promise greater returns with lower risk and investment than can a reuse strategy.

Because the use of learning object repositories is still at an early stage of development and not yet a proven method for reducing development costs, we did not consider the option of creating a true learning object economy involving payback to originators of materials. Instead, we focused on creating the conditions in which a repository system whose content would be free to potential users would work. Given the relative dearth of large repositories for e-Learning at the time of our research, one strategy we used was to examine reuse markets and repository mechanisms outside e-Learning to identify potential insights on successful reuse strategies that might apply to e-Learning. We found that these more mature markets for digital content have had mixed success with reuse and that they point to factors critical to success. For example, we found that success requires a relatively large potential market for reuse in order to generate a payoff that warrants investment. Further, some successful markets, such as the multibillion-dollar commercial Web-based visual and audio programming industry, suggest that even if demand is nominally present, one must have a high-quality product to attract a large consumer base. Other markets or instances of repository reuse, both commercial and government, have seen much more limited use, at least in part because of high transaction costs. For example, information in the much more modestly used Defense Audiovisual Information System (DAVIS) is relatively difficult to access and, once accessed, is difficult to customize.

The large commercial software industry, of which e-Learning is only a small part, has long attempted to foster reuse and can provide insights on how to develop conditions for creation of a learning object economy. Reusable software content can take many forms, including subroutines, functions, macros, libraries, objects, and design patterns. Whereas there have been notable successes in reusing software, achieving a positive ROI from reuse has been the exception in the software development industry and has proven difficult to predict. For example, one significant stumbling block to creating more-general software with a wide market for reuse has been the corresponding need for reusers to more heavily customize the output to fit their particular situations. The greater the cost of customization, the less economically viable the strategy of reuse. Areas in which software can be general enough to have a wide market for reuse while, at the same time, requiring minimal customization for most reusers have been discovered, but these areas—known as "sweet spots"—have not been numerous.

Another obstacle has been the multiple ways in which content can be organized, or "factored," to achieve the end goal. For example, software can be designed by dividing material by order of execution (phases), type of data, type of operation, or "tier." Having a similar factorization is important for reuse, because changing the factorization of otherwise appropriate content typically makes the prospect of reuse cost prohibitive. In e-Learning, the need for customization and the challenges involved in factorization are likely to be even greater than in the general software market. E-Learning embeds not just functionality, but also terminology, semantics, world view, pedagogy, subject matter, disciplinary context, and numerous other elements that may be crucial to the effectiveness of training and learning. Potential users of e-Learning software have expressed an especially great need to customize its capabilities.

The results from early experiences and the challenges likely to occur in the future both suggest that the success of an effective distribution system for learning objects will depend critically on the extent to which TD organizations are convinced of its value and the degree to which early adopters of reuse are able to realize and report positive returns. Thus, we recommend that ADL make ROI from reuse a specific area of near-term focus, comparing the current and prospective returns and risks of this strategy with those of other available options for reducing production costs. Further, we recommend that this focus be at the forefront of ADL's efforts to support e-Learning reuse, since other measures will make little sense unless a positive outlook for potential payback can be established.

Besides conducting research, ADL might employ strategies to foster the success of and positive perceptions of reuse. First, to build the economic case for reuse, ADL should seek to broaden the definition of reuse and document payoffs based on that wider view. This would mean supporting all five reuse approaches identified in our study, including recognizing and measuring concept reuse and structural reuse, as defined above.

Second, ADL can directly support organizations that are considering a reuse strategy for e-Learning by helping them learn to selectively design for reuse. Reuse experience in areas outside of e-Learning suggests that factors associated with a higher probability of success and reuse sweet spots include the presence of a big potential market for future reuse, the feasibility of reuse within and among organizations, and the potential for resolution of factorization issues. Acceptable balances between generalizability of content and the ability to customize and between high quality and low transaction costs are two additional factors.

Third, ADL might invest in high-profile pilots that illustrate the critical factors for achieving a positive ROI for learning object reuse. For example, ongoing efforts in military acquisition, medical training, or other areas might lead to opportunities for research measuring the ROI in promising areas. ADL might also promote research on reuse-related ROI by developing additional survey data and metrics within the planned DoD-wide registry for e-Learning content—the Advanced Distributed Learning Registry (ADL-R).

Disincentives to Sharing Are Currently a Secondary Challenge to Reuse but Could Threaten Future Successes

We also examined potential disincentives to reuse that may arise within organizations through stakeholders' reluctance to share learning objects or to reuse content created by others. Although many of the TD organizations we interviewed noted some reluctance to reuse among particular stakeholder groups, these tendencies were typically not cited as a critical factor impeding development of a reuse initiative. For example, only one TD organization cited disincentives as the "greatest obstacle" to reuse. Moreover, the most commonly cited disincentive among stakeholders, "Do not see significant benefits in reuse," appeared to be closely related to the ROI issue discussed above, and applied to both production of content and reuse of others' content. Another disincentive to designing for reuse among organizations' developers was the significant work involved (e.g., in producing metadata) that would be uncompensated and potentially at the expense of current customers (e.g., if they had to wait longer for products).

Finally, a moderate number of organizations noted that while custom content developers hired by TD organizations were currently cooperative and occasionally proactive, if reuse were to expand significantly, developers would lack sufficient incentives to comply with the "spirit of reuse"—i.e., to produce a sufficient amount of highly reusable content—because they would not accrue the profit from others' use of the content they created.

We expect that if large repositories become more prevalent and reuse becomes more common, disincentives to sharing content and reusing the content of others may go from being secondary obstacles to being a more significant problem. Research in knowledge management (i.e., the processes that organizations use to manage their intellectual assets) provides the foundation for this expectation. In addition to identifying obstacles to sharing information, as noted above, this literature shows that individuals and organizations are sometimes reluctant to use the knowledge of others. The predominant reason cited in our interviews for this reluctance was the effort required to revise content before it can be used for new purposes.

Other research shows that reluctance to reuse can also stem from concerns about the reliability of the material borrowed, fears of losing credit for ideas, and fear of becoming expendable. These issues were not of great concern among our interview respondents, who typically did not consider their training materials to be valuable intellectual assets. However, as the demand for learning objects grows, these incentive issues may become more prevalent.

Creating incentive mechanisms to counter stakeholder reluctance is key to motivating desirable behavior within large organizations. A variety of strategies might be used to create incentives for reuse, including measuring and rewarding the sharing and use of content, cultivating an organizational culture that favors reuse, assigning roles or providing support in a way that promotes reuse, tailoring technical systems to enable reuse, and using mandates or financial pressure to stimulate reuse.

One mechanism for addressing incentive issues is the high-level directive that requires reuse efforts within DoD: "Development, Management, and Delivery of Distributed Learning," DoD Instruction 1322.26, June 16, 2006. However, although this mandate will undoubtedly lead to larger DoD repositories of content and will nominally provide increased opportunities for reuse, it is unlikely to work well by itself. In addition to not addressing the potential disincentives discussed above, it may introduce further challenges. For example, two elements—the lack of general knowledge on how to design for reuse and the ease of complying with the "letter" but not the "spirit" of the directive—may lead to the flooding of repositories with content of little potential for reuse, thereby increasing the difficulty of finding truly reusable material. This could damage the perceived value of the emerging ADL-R at a time when supporting positive perceptions is critical to the success of an emerging learning object economy. Thus, supporting initiatives aimed at creating appropriate incentives may well be needed for the DoD directive to succeed.

ADL might pursue various options for stimulating additional incentive mechanisms for sharing content. Our earlier recommendation—that ADL work to raise the visibility of the potential ROI from reuse—applies to the entire area of incentives, not just to those relating to financial return. In addition, ADL might play an educational role by supplying TD organizations with information on how to foster positive organizational values among employees and providing training on how to design for reuse. Development of recognition systems, monetary and otherwise, might also help to encourage reuse. Furthermore, prior to the emergence of alternative business models, ADL might pursue more buy-in from custom content developers through appropriate policy and contract changes. For example, ADL might allow developer identification and contact information to be used in metadata so that highly reusable repository postings can serve as advertising and marketing tools. Finally, an investigation of incentive mechanisms appropriate to TD organizations could be pursued in the pilot demonstrations suggested above.

Organizational Processes for Implementing Widespread Reuse Will Need Extensive Development, Starting with Strategic Planning

Implementation of an RLO-based reuse strategy within a TD organization can require significant internal changes (e.g., with regard to the instructional design approach, business model, degree of collaboration with other organizations, use of technology, and other processes). Our interviews identified implementation issues as the greatest obstacle to overcome—much more important, for example, than issues related to technology or e-Learning standards. For organizations we interviewed whose reuse efforts had stalled or been abandoned, the need for strategic planning and increased collaboration were the most significant obstacles; for organizations that had had some success with reuse, metadata and repositories were the most notable obstacles.

These results suggest that organizations face a progression of challenges based on how far they have come in the change-management process. Some organizations are stuck at the beginning of the process (establishing a strategic plan), whereas others are more focused on how to implement an established plan. In the future, after large public repositories have been established, other obstacles might become prominent. What is clear from our study is that despite several years of effort, organizations are relatively early in the process of change, and much more development needs to occur.

Organizational experts note that effective implementation of a reuse strategy demands a supportive environment. Within TD organizations, support personnel are needed to facilitate collaboration between subject matter experts and technical staff. Internal guidelines for effective reuse are also important. Across TD organizations, the formation of repository communities can greatly increase the potential for reuse.

Such communities will need subject-specific standards and guidelines for reuse processes, along with a common language for metadata and metrics for measuring success. Guidelines that help organizations co-produce courses will also be needed.

Successful collaboration is also required for reuse success, whether the goal is to co-produce courses or to form effective repository communities. Such collaboration can often require a significant time investment. Collaboration can be aided by mechanisms to ensure communication, information sharing, and the development of trust. Successful collaboration may also require mandates, when appropriate, as well as the promotion of culture/values, involvement of sponsors, and engagement of neutral third parties to facilitate communication and prevent or reduce conflict.

We recommend that ADL support processes for implementing reuse strategies. ADL might take the lead in facilitating the creation of new reuse communities and support research documenting lessons learned about how to effectively implement reuse. ADL might also provide consultants to assist organizations that want to collaborate to foster reuse strategies.

ADL Can Encourage the Reuse Option for Reducing Development Costs by Taking a Proactive Approach

ADL should aid the development of a viable market for reusing learning objects by focusing on a key enabler—the perceived value of reuse. ADL can provide this service in two key ways: by helping organizations engaged in making an initial decision about whether and when to invest in reuse determine the potential for reuse and the conditions leading to its greatest payoff, and by increasing its support to early adopters that have already begun implementing their reuse strategies. The recommended approaches, as described above, are as follows:

- Broaden definitions of reuse and redefine success through the use of metrics and surveys.

- Invest in high-profile pilot programs to identify conditions with the highest potential payoffs for reuse.
- Conduct or sponsor research to evolve guidelines for implementing reuse strategies.
- Evolve ADL's role as a neutral trusted advisor to TD organizations.

ADL might also sponsor research efforts to develop a better understanding of how reuse efforts can be supported. Possible projects are (a) an evaluation of approaches for improving search capabilities for digital training content; (b) development of additional metrics for ADL-R's scorecard to capture costs and benefits to both contributors and seekers of content; (c) an evaluation of the evolution of the "DL supply chain" over time in order to predict the interventions that could speed up the process to rapidly produce high-quality content at low cost; (d) focused case studies of current, high-profile efforts to maximize reuse of training content and document emerging lessons learned and sweet spots for different types of reuse; (e) development of guidelines and a decision tool to help project/program leaders determine the likelihood of successful reuse.

These actions may not be enough to create a viable market for reusable e-Learning content (e.g., cost savings may prove to be too small), but they may well be necessary if promising reuse efforts are to successfully emerge and realize their wider potential in the near term.

Acknowledgments

This study benefited from the assistance of many people in the DoD Advanced Distributed Learning (ADL) Initiative and affiliated organizations. We are particularly indebted to our sponsor, Robert Wisher, Director of the ADL Initiative, who provided his guidance, as well as his willingness to ask difficult questions, hear challenging answers in return, and share those findings with the broader, SCORM (Sharable Content Object Reference Model) community. We also wish to thank Joe Camacho, U.S. Joint Forces Command, and Dan Gardner, Office of the Secretary of Defense, for their strong roles in supporting this research.

We also received support and intellectual input from other members of the ADL community. This included conversations with and critiques from David Daly, Eric Roberts, Judy Brown, Paul Jesukiewicz, Dexter Fletcher, Jean Burmester, and Rebecca Bodrero. Special thanks go to Avron Barr for insights provided during stimulating discussions.

Our research required input from many members of different education and technology communities. We were fortunate to be connected with Gerry Hanley of MERLOT (Multimedia Educational Resource for Learning and Online Teaching), David McArthur at the National Science Foundation, and Wayne Gafford, now with ADL, all of whom provided information and valuable pointers to resources early in the project. Insights into the history of efforts to implement standards for learning objects and develop related "markets" came from our conversations with Jim Spohrer and Chuck Barritt. Additional thanks go to Michael Parmentier and Clark Christiansen of Booz Allen Hamilton.

Two people who served as particularly valuable sources of information and nexuses of contacts and to whom this work owes special debts are Robby Robson of Eduworks Corporation and Elif Trondsen of SRI Business Intelligence.

Two consultants to the project, Joshua Sharfman and Bryan Chapman, deserve thanks for bringing their insights and resources from the commercial world of digital content management. Sharfman shared his real-world experience in developing and fielding large-scale, complex repositories; Chapman, of Chapman Alliance, provided comments on early ideas and helped us track down key literature and contacts relevant to our analyses.

We are also indebted to several RAND colleagues who contributed to the project: Danielle Varda, Susan Gates, Jim Hosek, Naveen Mandava, and the internal RAND reviewers of an earlier version of this report, Bob Anderson and Christopher Paul.

Finally, we would like to extend our sincere thanks to those from distributed learning publishing organizations who participated in extended interviews and telephone surveys concerning their experience with e-Learning reuse. Without their willingness to speak frankly and trust our team to uphold promised standards of anonymity, important components of our analyses would not have been possible. Because of our promise to assure confidentiality, we cannot thank these individuals by name, but we want them to know that their willingness to share their own experiences and insights for improvement were very important to completing our analyses and producing viable recommendations.

Abbreviations

3D	three-dimensional
ADL	Advanced Distributed Learning (Initiative)
ADL-R	Advanced Distributed Learning Registry
AMEDDCS	Army Medical Department Center and School
CD-ROM	compact disk, read-only memory
CMS	content management system
COTS	commercial off the shelf
CSREES	Cooperative State Research, Education, and Extension Service
DARPA	Defense Advanced Research Projects Agency
DAU	Defense Acquisition University
DAVIS	Defense Audiovisual Information System
DITIS	Defense Instructional Technology Information System
DL	distributed learning
DoD	Department of Defense
e-Learning	electronic learning
ERP	Enterprise Resource Planning
FY	fiscal year
IMI	interactive multimedia instruction
IT	information technology

JKDDC	Joint Knowledge Development and Distribution Capability
JSF	Joint Strike Fighter
KM	knowledge management
KMS	knowledge management system
LCMS	learning content management system
LMS	learning management system
MERLOT	Multimedia Educational Resource for Learning and Online Teaching
NATO	North Atlantic Treaty Organization
NAVAIR	Naval Aviation Systems Command
NPGS	Naval Post Graduate School
NRO	National Reconnaissance Office
PDA	personal digital assistant
PEO STRI	Program Executive Office for Simulation, Training and Instrumentation
RDECOM	Research, Development and Engineering Command
RLO	reusable learning object
ROI	return on investment
SCO	sharable content object
SCORM	Sharable Content Object Reference Model
SME	subject matter expert
TD	training development
TLO	terminal learning objective
UK	United Kingdom
USDA	United States Department of Agriculture
VA	Department of Veterans Affairs
VTT	video teletraining

CHAPTER ONE
Introduction

Reuse of Content Is a Strategy for Reducing the Development Costs of Distributed Learning

The Department of Defense (DoD) is interested in expanding the use of distributed learning (DL) for military training. DL not only offers learners the promise of greater flexibility by providing opportunities for self-paced learning and training at any time and in any place, but also increases the reach of training organizations by expanding the means and technologies available for developing and delivering training. DL also allows automated tracking of student performance and promises to help standardize course content.

Although demand for DL is increasing, DL still represents a small percentage of all learning.[1] An important impediment is the high cost of developing and maintaining electronic learning (e-Learning) content. For example, one survey in the commercial sector found that the average cost of developing interactive multimedia instruction (IMI), level 2–3,[2] in 2007 was $24,700 (Brandon Hall Research, 2007). The

---

[1] One industry group estimated that only 2 percent of the U.S. education and training market is in technology-based products (Ambient Insight, 2006).

[2] Level refers to "levels of interactivity" between the learner and DL media, with higher levels involving more learner participation. Levels of interactivity have universal meaning but can have somewhat different definitions in practice depending on industry and context. To cite one example of a definition, Level 2 interaction means that the learner has some control over lesson activities, such as the ability to click on icons to reveal information, to move objects on the screen, to fill in forms, and to answer questions. Level 3 interaction involves more participation, which comes from the use of scenarios for testing, the need

high costs of moving courses to DL are caused in large part by the need to develop content independently for each course.

Development costs might be reduced if digital learning content could be reused across courses on a large scale.[3] *Reuse* can be defined as the use of existing digital content to produce new content, or the application of existing content to a new context or setting. In the pre-digital world, a reuse analog might be the use of existing textbooks on a subject as a starting point for designing a new textbook, as opposed to designing a new textbook from scratch. Just as libraries and bookstores provide a distribution mechanism in the non-digital world, widespread digital reuse will require a mechanism to bring together reuse "buyers" (e.g., trainers, end users) and "sellers" (e.g., training development [TD] organizations, authors). The main products in this environment would be reusable learning objects (RLOs)—chunks or modules of digital learning material that can be stored in searchable databases (learning object repositories) and then accessed by third parties to create new course content.

Research Questions Examined

In 2006, RAND was asked to examine how the Advanced Distributed Learning (ADL) Initiative and DoD more broadly might encourage reuse and support the development of a successful distribution mechanism for reuse efforts. In particular, the study focused on the extent to which incentives (both economic and non-economic) and other enablers currently are, or might be, used to encourage TD organizations to develop the supply side of a distribution mechanism for reuse.

---

for the learner to make decisions, and the use of complex branching based on the learner's responses.

[3] Another potential advantage of reuse is improved quality and validity of course content, because instructional material is developed with input from a greater number of experts and users and because there are increased opportunities for vetting content across courses.

Four key questions guided the research, emphasizing return on investment (ROI), potential disincentives to reuse, and implementation of a reuse strategy:

1. To what extent are TD organizations currently engaged in reuse at this stage of technological development?
2. To what extent will organizations find reuse a worthwhile investment?
3. To what extent will disincentives to wider sharing of learning objects impede reuse?
4. To what extent will organizations know how to implement a reuse strategy?

Our decision to focus on incentives and implementation was based on the recognition that many innovations fail not because of their technical merits, but because of problems in implementation.

Preview of Key Findings and Recommendations

RLO-based reuse is relatively rare, and technical challenges will take time to overcome. We found relatively little current use of RLOs. Although our interviews with TD and other organizations indicated that RLOs are being produced and that technical standards for sharing RLOs are well established, the adoption of RLOs is in the early stages, and improvements in interoperability are still needed. Authoring technologies and content management systems (CMSs) are not yet up to expectations.

Significant returns are the exception, and future successes remain difficult to predict. Training organizations are typically not getting much of an ROI for reuse. Savings are typically modest and not derived from the development of RLO repositories. Our research suggests that in the future, the ROI for reuse is not likely to be balanced across investments, and that the potential for high ROIs depends on organizations identifying "sweet spots," or areas in which software can be general enough to have a wide market for reuse and yet require minimal customization for most reusers.

Disincentives to sharing are currently a secondary issue but could threaten what could otherwise be future successes. Current barriers noted by TD organizations included the lack of perceived benefits and the significant amount of work required by TD organizations, with little promise of ROI in the near term. However, if reuse becomes more prevalent in the future, disincentives to knowledge sharing may become more prominent— for example, because of concerns about free riding or ambiguity about the reliability of the knowledge provided in RLO repositories.

Processes to advance widespread reuse need extensive development, starting with strategic planning at the organizational level. Implementation of a reuse strategy will require significant change management for TD organizations, including potential changes in the instruction design approach, business model, and collaboration techniques. Collaboration across organizations could become especially challenging given the investments of time and resources that would be needed.

The ADL Initiative could encourage reuse by making the determination of ROI an organizational focus and by evolving its role as trusted advisor on reuse implementation issues. The role of reuse and RLOs can be expanded, but success will require that the emphasis be on opportunities with high potential payoffs. ADL can help to support the perceived value of reuse in an emerging learning object economy by determining the true potential for reuse and identifying the conditions that lead to the greatest payoff. To support early adopters of a reuse strategy, ADL might focus on evolving its role as trusted advisor to TD organizations and on increasing the support it provides to potential reusers by developing guidelines for reuse and disseminating best practices.

Focus of Project: Supply Side and Market Enablers

Figure 1.1 shows how suppliers and demanders might interact with repositories of e-Learning content. While the illustration on which this figure is based used the word *marketplace* (Johnson, 2002) to describe

Figure 1.1
Elements of an Exchange System for Reusable Learning Objects

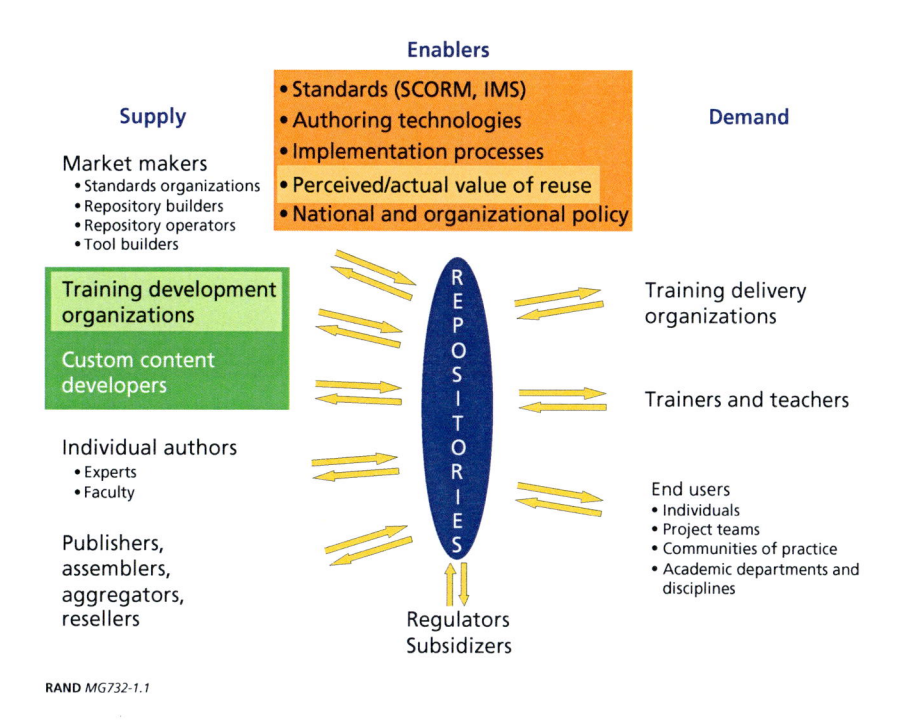

Enablers
- Standards (SCORM, IMS)
- Authoring technologies
- Implementation processes
- Perceived/actual value of reuse
- National and organizational policy

Supply

Market makers
- Standards organizations
- Repository builders
- Repository operators
- Tool builders

Training development organizations

Custom content developers

Individual authors
- Experts
- Faculty

Publishers, assemblers, aggregators, resellers

REPOSITORIES

Regulators
Subsidizers

Demand

Training delivery organizations

Trainers and teachers

End users
- Individuals
- Project teams
- Communities of practice
- Academic departments and disciplines

RAND MG732-1.1

the environment, the current DoD context does not provide for the buying or selling of content; instead, suppliers are required to place content in repositories from which qualified demanders can access it for free. On the supply side are TD organizations, authors, publishers, market makers, and others with a requirement to bring RLOs into a repository. On the demand side are training delivery organizations, trainers, and end users, all of which could potentially benefit from the content of the repository. Use of the repository is facilitated by key enablers, including standards, authoring technologies, policies, and the perceived value of reuse.

Such a mechanism for exchange exists only in limited form today, and repositories of RLOs are typically internal to a particular organization and vertical in structure (rather than cutting across various organizations). For example, the U.S. Army has its own environment for reuse, which involves internal TD organizations, trainers, and others. While these internal repositories can be useful, they do not allow for reuse across organizations, and they typically leave out independent actors (e.g., training developers, authors, publishers).

Our research considered ways to encourage the development of a broader learning object economy that cuts across organizations and establishes incentives for participation. For this project, we focused mainly on ways in which TD organizations might be encouraged to develop the supply side of the reuse "market" when no compensation is involved. We also focused mostly, though not exclusively, on defense organizations, since DoD has the greatest leverage with these organizations, and a mandate for a change toward reuse already exists.

Implementing a strategy for reuse would require changes and additions to the supply side of the e-Learning repository system. For example, suppliers in many cases would need to think differently about the product they are developing. The traditional model of training development has centered on the course and the various threads connecting different parts of the course. In contrast, a reuse model requires developers to think in terms of learning objects—components or modules of instructional material that are not tied to an individual course but, instead, can be combined to fit the needs of many courses and instructional practices. This shift to a learning object model represents a new way of doing business for training developers.

Incentives and other enablers would also have to evolve. One of the major themes of this research is that TD organizations will not develop RLOs unless decisionmakers can perceive the *value* of reuse. As will be discussed later in this monograph, key personnel in TD organizations presently do not perceive the value of investments in reuse.

Focus of Research Approach: Early Adopters of Reuse for e-Learning and Reuse in Other Contexts

To answer our research questions, we conducted structured telephone interviews with individuals at large TD organizations that had all pursued a reuse strategy for e-Learning. These organizations covered a wide range of pursuits and were located in DoD and foreign defense organizations; other, non-defense U.S. government organizations; the commercial sector; and academia. We also conducted site visits and more-extensive interviews with training developers and other stakeholders in two of the defense-related organizations.

Our sample of organizations was not random. We focused on large organizations that were early adopters of a reuse approach and whose strategy involved multiple stakeholders and internal repositories of e-Learning content. Candidate organizations were identified in various ways: through discussions at conferences dealing with learning standards, reuse, or DL; by searching the literature and Internet for these same subjects; and by asking respondents known to pursue reuse and other ADL experts about organizations they knew had invested in a reuse strategy. In seeking interview candidates, we were careful to specify that we were interested not only in organizations that had experienced some degree of success with reuse, but also in those that had experienced failure or had decided to abandon a significant initial effort. We attempted to contact about twice as many organizations as eventually agreed to participate. Only two organizations declined to participate after we had made contact and explained the project's objectives.

Interviews with people at the TD organizations were conducted using a semi-structured interview format (see Appendix B for the interview and protocol) over a period of 60 to 90 minutes. The interview consisted of a list of open-ended questions that were posed to all participants. Responses were later coded according to categories developed by the research team to cover the types and range of responses received.

Participants in each interview were selected by our main point of contact for the particular organization; they included training department heads, program managers, instructional designers, technical experts, subject matter experts (SMEs), and custom content contractors working for the organization.

We also conducted a large number of interviews with ADL experts in various specialties (e.g., e-Learning standards, custom content development, digital rights management). We did not use a formal interview instrument in these cases. Instead, we conducted more-informal discussions on the enablers of and obstacles to reuse, as well as on potential improvements in the implementation of reuse strategies.

Because the size of existing defense-related learning repositories is limited, we also looked at reuse markets outside e-Learning (i.e., software and materiel development) for insights on successful reuse strategies. Further, we reviewed the literature on knowledge management (KM), particularly on incentives and disincentives surrounding the sharing of knowledge. Finally, we conducted interviews with experts involved in reuse within other marketplaces for digital objects.

Table 1.1 lists the major defense-related and other TD organizations that took part in our formal interviews.

Interviews concentrated on representatives of DoD and the defense industry, although we also spoke with representatives from commercial and academic organizations to gain a broader view of the potential for reuse.[4] Because the interviews promised confidentiality to all interview participants, no specific names are mentioned in this monograph in connection with any findings.

Organization of This Monograph

The remainder of this document is divided into five chapters: Chapter Two examines the prevalence of reuse and the role of standards and technologies; Chapter Three looks at economic incentives for reuse; Chapter Four focuses on disincentives to sharing knowledge; Chapter Five discusses issues related to the implementation of reuse approaches; Chapter Six provides overall recommendations.

---

[4] Based on the nature of content alone, the academic environment provides greater potential for reuse than the military environment does. The academic environment comprises primary and secondary schools and universities, which together represent a substantially larger group of learners than those in the military environment. Also, while academia teaches essentially the same subject matter within a great number of schools and to a large number of students, much of military training is contextual (e.g., training to perform specific tasks in specific situations) or related to many different pieces of equipment, any one of which is used by only a small percentage of the military.

Table 1.1
Organizations Participating in Formal Telephone Interviews

DoD and international defense
    Army
    Navy
    Naval Aviation Systems Command (NAVAIR)
    Joint Strike Fighter (JSF)
    National Reconnaissance Office (NRO)
    Defense Acquisition University (DAU)
    Naval Post Graduate School (NPGS)
    Army Medical Department Center and School (AMEDDCS)
    Joint Knowledge Development and Distribution Capability (JKDDC)
    NATO
    UK Defense
Other government
    Internal Revenue Service
    Department of Veterans Affairs (VA)
Commercial
    Boeing
    Cisco
    Apple Inc
    Northwest Airlines
    Verizon Wireless
    DaimlerChrysler
    Tweeter Home Entertainment
Academic
    Johns Hopkins University
    Korean education
    Global Health Network University

CHAPTER TWO
The Prevalence of Reuse and the Role of Standards and Technologies

This chapter focuses on the nature and extent of current reuse strategies, as well as on the importance of current reuse standards and technologies in implementing those strategies. Although standards and technologies were not a primary focus of our research, they are discussed here to provide perspective on our findings regarding ROI, incentives, and implementation, which are discussed in later chapters. The main points here are that RLO-based reuse is relatively rare, that this rarity is partially explained by the only recent emergence of reuse standards and technologies, and that greater maturity in both of these areas will take time to achieve.

Organizations Pursue Reuse Using a Number of Approaches

Our early inquiries determined that TD organizations have primarily used the following three approaches in pursuing a strategy of reuse (either within their own organization or across related organizations):

- a *top-down, coordination-driven* approach, in which TD organizations collaborate on the design or otherwise coordinate so that e-Learning courses can reach wider audiences
- a *reusable learning object (RLO)* approach, in which TD organizations design and reuse digital content as independent objects, complete with learning objective(s), interaction, and assessment

- a *bottom-up, asset-driven* approach, in which digital assets (e.g., images, sound, video) are reused.

These approaches to reuse differ in terms of the granularity, or *grain size*, of what is being reused (e.g., entire course content, versus RLOs, versus digital assets, such as images), the existence and maturity of markets involving the approach, and ways of implementing the approach. With regard to implementation, the types of reuse differ in terms of personnel involved (and whether they are known or unknown to each other), level of collaboration required, technologies employed, and types of internal change processes likely needed as prerequisites.

The Top-Down Approach

The simplest form of the top-down reuse approach occurs when an entire course is redeployed to new student populations. A commercial market for such redeployment exists for training of large target audiences (e.g., information technology [IT] training, leader training, language training). Suppliers of this sort of training are known as commercial, off-the-shelf (COTS) content developers. At the other end of the reuse scale, redeployment can sometimes mean the migration of a course to one additional institution or student base.

Another form of top-down reuse occurs when content is purposefully employed in so-called multiple-use cases—e.g., when digital content is designed for more purposes than just training, such as technical documentation or use within a help system. Still another form occurs when organizations seek to reuse content in multiple forms of delivery—i.e., when training content is used in various formats, such as for presentation on a laptop and a PDA (personal digital assistant). Finally, we include in our definition of top-down reuse the case in which organizations co-design e-Learning so that it fits the needs of each organization when completed. In this case, organizations do not implement a reuse strategy as much as they "plan for greater use."

Top-down reuse saves resources by eliminating duplicated efforts in development and in strategic planning. Often this form of reuse applies to individual courses and not to an organization-wide initiative. Top-down reuse does not always involve collaboration among organizations, but it almost always requires interoperability of training content. Top-down approaches do not always reduce development cost, but they do increase the number of people who can take advantage of a course or course content once it has been developed. In other words, top-down reuse reduces the per-student cost but does not necessarily reduce the per-course cost.

The RLO Approach
The RLO approach to reuse involves the design of digital e-Learning content in portions smaller than courses, or "chunks."[1] For example, an RLO might be created for each lesson or learning objective. In addition to content, an RLO is typically characterized by interaction between the learner and the course, including a test to verify success. One way to think of an RLO is as a compromise between the top-down and bottom-up approaches. Instructional designers and administrators tend to focus on the course (or at least the lesson level) and make all content self-contained, whereas technical developers tend to focus on the digital assets. As a compromise, RLOs seek the smallest learning chunks that are self-contained and have an identifiable objective.

By focusing at the level of the learning object, the RLO approach provides many more opportunities for reuse than courses do and promises to reduce the collaboration costs required for reusing an entire course. However, this approach also implies a greater management burden related to storing and describing many more objects in a way that allows later retrieval. Moreover, the RLO approach raises instructional design and technical issues because an object's reusability will depend on the extent to which its content can stand alone and be kept separate from context, pedagogy, structure, and presentation.[2] Thus, the RLO approach requires the creation of new instructional design processes, repositories, metadata systems, search tools, and authoring tools.

---

[1]   For a complete description of this approach, see Barritt and Alderman, 2004.

[2]   Robby Robson, Eduworks Corporation, in a teleconference interview conducted by Michael Shanley and Matthew Lewis, two co-authors of this monograph, on August 4, 2006.

The Bottom-Up Approach

Bottom-up reuse entails sharing digital assets in multiple contexts. Assets can include images, audio, video, PowerPoint files, animations, or other digital material—all with no reference to context, such as instructional methods, skill hierarchies, or learning objectives. No RLOs are required. Asset reuse still requires labeling, with good metadata, storage in repositories, and search tools. However, unlike the RLO approach, no changes to development processes are required to design an asset's specifics into new instructional content. Instead, the asset is specifically chosen for the context in which it will be included. Since it is free of such context, it requires no labor or tools to extract the asset from context for repurposing. Commercial markets for digital assets are well developed in some areas, including photography, music, sound effects, and three-dimensional (3D) models.

Processes in a Reuse Strategy

In addition to observing differences in the grain size of reusable content, we observed differences in approaches to the process of reuse. One process is strategic, for specific audiences. Organizations that use this process collaborate for the purpose of designing a course to meet the needs of two or more specific student populations. A second approach to the reuse process is also strategic, but in this case the audiences are not necessarily known. Here, one or more organizations design content to be generic, the intent being to make it apply to multiple (and perhaps unknown) student populations.

Finally, we observed a few instances of ad hoc, or unplanned, reuse, whereby reusable content was identified by members of organizations through social networks, database searches, and other informal means. For example, through informal social interactions, members of a network of organizations discover that some of their courses are of interest to multiple populations of students. Subsequently, the main organization sends its courses to other organizations or redeploys other organizations' courses on its own learning management system (LMS).[3]

---

[3] An LMS is software that automates the administration of training from multiple courses, which can come from a variety of sources. For example, an LMS is designed to register users,

Technically, one might view this type of reuse as an anomaly to the approaches distinguished above in that it has both a high level of grain size and emanates from an uncoordinated and bottom-up source.

RLO-Based Reuse Is Less Prevalent Than Other Approaches

Although most organizations pursued multiple approaches to reuse, we found that the RLO approach was less prevalent than the others. Figure 2.1 shows the prevalence of the different strategies for reusing content among the training developers we interviewed.

While most organizations produced RLO content, and most had three or more years of experience pursuing an RLO-based reuse strat-

Figure 2.1
Extent to Which Different Approaches to Reuse Have Been Employed

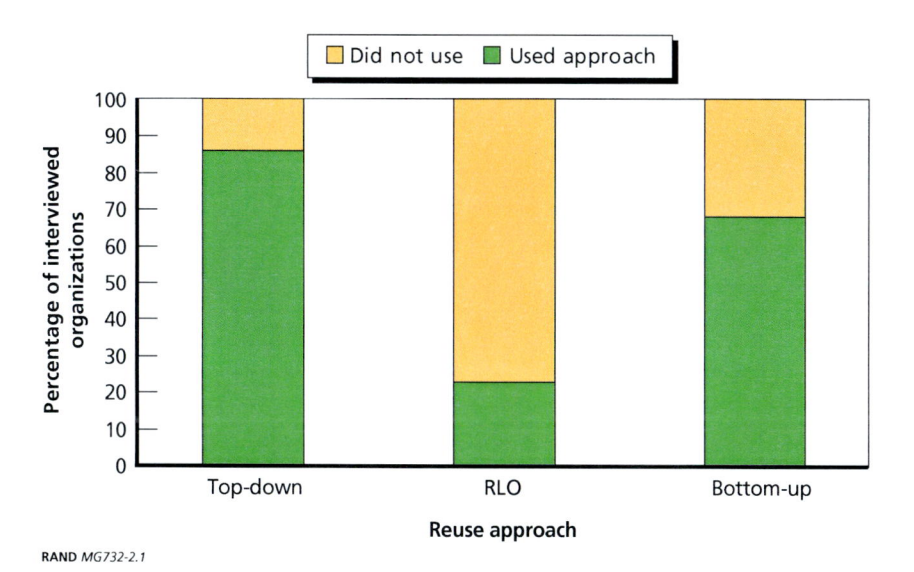

---

track courses in a catalog, record data from learners (e.g., test results), and provide reports to management on administrative activities. It usually does not include its own authoring capabilities; instead, it focuses on managing courses created by a variety of other sources.

egy, only a little more than 20 percent of them reported successful reuse with the RLO approach. This seems a particularly small number given that we sought out organizations with the greatest reuse experience.

In contrast, 70 percent of the TD organizations reported employing the bottom-up, asset-driven approach. Moreover, 85 percent of the organizations used some form of the top-down, coordination-driven approach. The implementation for some of these efforts involved sophisticated collaboration, but the most prevalent form of reuse was simple redeployment of entire courses.

It is likely that the importance of the RLO approach will not change soon. Sales of learning content management systems (LCMSs)[4] and industry information suggest that few organizations are actively pursuing reuse.[5]

However, aspects of the top-down and bottom-up approaches to reuse in some ways mirror what is needed for an RLO approach and could eventually lead to greater RLO reuse. For example, the bottom-up strategy involves contributions to and use of repositories in the development of training content, and the top-down strategy tends to involve collaboration and formation of partnerships, both of which are necessary for successful implementation of an RLO reuse strategy (see Chapter Five for further discussion). Further, it is worth noting that most of the TD organizations we interviewed expressed optimism about the future of a reuse strategy, despite the limited success of reuse to date.[6]

---

[4] An LCMS is software that provides an authoring application, a data repository, a delivery interface, and a variety of administration tools to aid in the management of e-Learning content. An LCMS allows users to create and reuse digital learning assets and content within a common authoring environment.

[5] See Brandon Hall Research's *LMS Knowledge Base, 2005–6* and its *LCMS Knowledge Base* (Brandon Hall Research, 2006c and 2006b, respectively). The LMS study accounts for 19,417 implementations of LMSs, but the LCMS study shows only 1,887 implementations of LCMSs. Although the two reports do not cover exactly the same technologies, there is a significant overlap in vendors that have both an LMS and an LCMS. From this information, it is apparent that an LCMS is not always a standard fixture when an LMS is installed. For example, in the case of a vendor that sells an LMS with an LCMS built in, only 150 of its 460 customers used their LCMS module.

[6] We often speak of the view of a *TD organization* or a *respondent*, by which we mean, in both cases, the majority view among those we interviewed that are associated with that

On a scale of 1 to 10, with 10 being the most optimistic, the average response was 8. Organizations saw much of reuse's potential as not yet having been tapped, both in house and in similar organizations.

Relatively New Technical Standards and Technologies for Reuse Partially Explain the Relative Scarcity of the RLO Approach

According to some interview participants, one reason for low use of the RLO approach is that although technical standards for sharing content are well established, adoption of these standards is still incomplete and improvements in interoperability are still needed. Moreover, even though authoring technologies and CMSs in support of reuse are seen to be evolving, they are not, according to some interview participants, to the point of being user friendly and cost-effective enough to meet organizations' needs.

Interview Results: Technical Standards and the Sharable Content Object Reference Model (SCORM)[7]

- Nearly all (90 percent) of the organizations used SCORM, although many did not use the latest version.
- Existing standards (and supporting authoring tools) were seen as reuse enablers: Sixty-seven percent of the TD organizations found them helpful or critical to success; 33 percent found them helpful but thought they could provide more.

---

organization. In other cases (see, for instance, Chapter Four), we speak about stakeholders' viewpoints, by which we mean the views of subgroups associated with TD organizations that are involved with the TD process—for example, content developers or program managers. Views attributed to a *representative* of an organization or an interview *participant* are those of an individual.

[7] SCORM is a standard for a technical framework to enable the use of Web-based e-Learning content across multiple environments (e.g., LMSs). SCORM defines how individual instruction elements are combined at a technical level and sets conditions for the software needed to use the content. For further explanation, see "SCORM 2004, 3rd Edition" (Advanced Distributed Learning, 2008).

- About 40 percent felt that standards posed a current obstacle, though none saw them as the greatest obstacle to reuse. Among the obstacles listed as examples (see Appendix B), "standards" had the fewest number of hits.
- Interoperability has not been fully achieved. Organizations reported that in nearly every attempt to develop reusable content, a course required technical adjustments to run on an LMS other than the one for which it was created, even if both systems were certified as SCORM conformant. Further, a few organizations reported that trying to convert to the latest version of SCORM had become a significant distraction within the organization. Thus, standards in e-Learning are helping but have yet to become "invisible," as good standards are.
- A number of organizations felt that SCORM limited the use of the organization's desired instructional strategy.

Interview Results: Technologies That Support Reuse

- About 50 percent of the TD organizations felt that technologies pose a current obstacle to reuse, though only one saw them as the greatest obstacle.
- A number of organizations reported that LCMS capabilities have not matched their expectations. They reported that the tools took a long time to integrate into their organizational processes (e.g., months were required for training and conversion) or were not user friendly in fulfilling all the organization's needs.
- Other organizations stated that authoring tools to support reuse were generally not well developed, did not have high capability, were not contained in integrated packages, or were not compatible with each other.
- Some felt that it was easier to create new content from scratch than to repurpose existing content.

In general, the interview responses suggest that technologies supportive of reuse are in an early stage of development. The implications of this finding can be seen in the technology adoption life cycle shown in Figure 2.2.[8]

According to this life cycle, new technologies generally have a small group of initial "innovators," or people who try them out and/or experiment with them. If a technology shows promise and provides results, "early adopters" then take it on and wait for an assessment of its viability. However, between the early adopters and the next group, the "early majority," there is a "chasm" that must be filled with a critical mass of less-technologically-savvy users who are willing to commit to using the new technology. Crossing this chasm (represented by a space between adjacent groups of users in Figure 2.2) places a large

Figure 2.2
Place of RLO-Based Reuse in the Technology Adoption Life Cycle

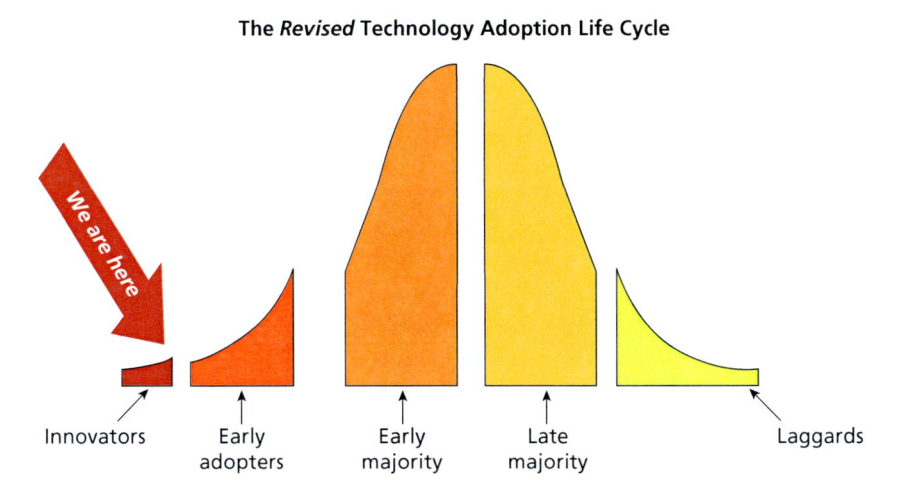

The *Revised* Technology Adoption Life Cycle

Innovators — Early adopters — Early majority — Late majority — Laggards

We are here

SOURCE: Moore, 2002.
RAND MG732-2.2

---

[8] This life cycle is adopted from a revised edition of Geoffrey Moore's 1991 *Crossing the Chasm: Marketing and Selling High-Tech Products to Mainstream Customers* (Moore, 2002). Dr. Tom Byers, Faculty Director of the Stanford Technology Ventures Program, describes Moore's work on technology development and adoption as "still the bible for entrepreneurial marketing 15 years later." (Quotation is from material by and about Tom Byers available at ecorner, Stanford University's Entrepreneurship Corner. See Byers, undated.)

burden on developers to provide usability and technical support for the technology.

Technologies that support reuse are still in the earliest stages of this cycle and still face serious challenges to wider adoption. Progress toward widespread adoption is likely to be relatively slow for many reasons, including the normal organizational "costs" associated with adoption of any new technology and the need for changes in related development processes. In addition, the lack of demonstrated ROI for such technologies will slow their adoption (and note the relatively slow adoption of LCMSs that we mentioned above). For example, some of the early tools are currently government funded and thus not yet in commercial markets. The viability of emerging tools for cost-effectively meeting the reuse needs of TD organizations will be explored in the coming years. Whether the potential benefits of using such tools will outweigh the earlier-mentioned costs and lead to market acceptance—and to crossing the chasm—is yet to be determined.

Concept Reuse Avoids the Technical Problems of RLO-Based Reuse

Given the technical issues that reusers currently face, we identified a fourth important type of potential reuse: concept reuse. We believe this more traditional approach tends to be overlooked in the presence of emerging technologies. Concept reuse is the employment of pedagogical approaches from other courses, including instructional methods, task decompositions, and assessment methods. This practice is traditional in that it is parallel to a researcher's use of related research papers to design his or her project or the inspection and analysis of existing Web sites as models for the structure and content of new Web sites. Concept reuse saves design costs, which can be significant in e-Learning, but does not require interoperability or technologies that repurpose existing digital content.

It is essential that concept reuse be acknowledged so that it can be measured and documented as an early success in the effort to reuse, and so that it can be supported in the design of large-scale repositories. The success of concept reuse requires the ability to quickly locate target content and explore it for possible emulation or partial structural replication. This, in turn, requires the ability to quickly search for and access content or content summaries for inspection. These capabilities are not built into the design of early large-scale repositories. For example, the Advanced Distributed Learning Registry (ADL-R) presently allows potential reusers to access only metadata, along with a pointer to the repositories where the content resides. If concept reuse is to be supported, potential reusers may well need to access (or at least index) and search content or the underlying code that supports content. The capability for such searches exists today. For example, Google has a prototype search engine for computer code that is "public source"[9] to aid programmers looking for specific elements of computer code.

---

[9] Google Corporation has a "labs" set of public prototypes of tools, one of which searches for public-source computer code (Google, 2009).

CHAPTER THREE

Economic Incentives

This chapter considers the ROI that organizations have experienced in implementing their reuse strategy in the e-Learning context. While none of the organizations interviewed could cite specific ROI figures, most were comfortable making broad comparative statements about reuse. In addition, this chapter examines experience with reuse in a variety of contexts (e.g., e-Learning, distribution of other digital products, software in general, and materiel development) and the implications for DoD. Our goal was to identify factors that can explain success with a reuse strategy more generally and that can guide others in considering reuse designs for e-Learning in the future.

On the whole, our interviews suggest that significant returns with a reuse strategy in e-Learning are the exception. Even after several years of pursuing reuse, few TD organizations had more than modest returns on their investment in the e-Learning context. In the broader context, we have concluded that predicting or generalizing about what determines the success of a reuse strategy will be difficult. We recommend that ADL focus on obtaining additional information on ROI issues in e-Learning; we also discuss options for how ADL could pursue that goal.

Few Training Organizations Had More Than Modest Returns from Pursuing Reuse

What we learned in our interviews suggests that few early adopters of an RLO-based approach had more than modest returns on their investment. Some even claimed that their returns were negative. Although TD organizations do not typically have concrete measures of ROI, we asked them to estimate their returns in a broad way and to compare them with their expectations when they began pursuing reuse. Most of the organizations we interviewed had pursued reuse for three to six years and thus had enough experience upon which to base an estimate. Only two organizations were excluded from this analysis because of limited experience with reuse.

As Figure 3.1 shows, 35 percent of the interviewed organizations reported no savings at all or a net loss from their reuse efforts. All of these organizations were referring to attempts to implement an RLO-based reuse approach. Most often, this meant that an organization had made a significant investment in changing its processes and procedures but had not yet successfully reused content. A few organizations, some of which had sunk significant revenues in a reuse strategy, told us that they had abandoned reuse initiatives, concluding that they would

Figure 3.1
The Extent to Which Organizations Saved e-Learning Development Resources by Employing a Reuse Strategy

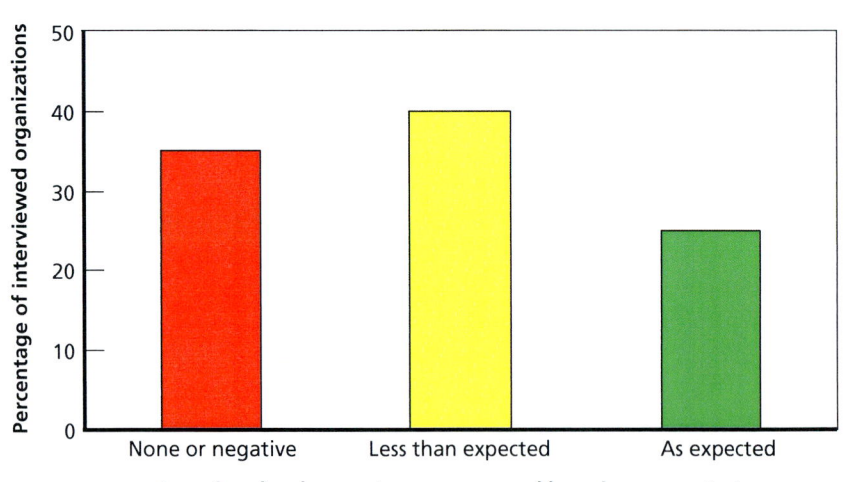

never recoup their investment and needed to cut their losses. In these cases, the organizations felt that the potential for reuse within their own organization had been greatly overestimated.

Forty percent of organizations estimated that they had achieved some returns but less than expected. Often, respondents felt that if any real gains had been made, they were no more than small. The reason for this result was more often than not that obstacles to reuse had turned out to be greater than expected (this point is discussed further in Chapters Four and Five).

Of the organizations represented by the first two bars in Figure 3.1, 60 percent (or 45 percent of all organizations) also listed "insufficient ROI" as an obstacle to moving forward with initiatives to increase RLO-related reuse. However, only those abandoning their reuse strategy saw this obstacle as the greatest one they faced.

Twenty-five percent of the organizations estimated a positive ROI that was in line with their pre-implementation expectations. These organizations typically employed either the top-down or the bottom-up approach to reuse, or both. Two respondents estimated rather modest savings, in the range of 5 to 15 percent. Two others remarked that the main benefit was not cost reduction but an ability to increase the amount of training delivered within a fixed budget.

Two respondents in Figure 3.1's "as expected" column estimated large cost savings (e.g., were able to cut the average cost of development in half). Those savings did not derive from reuse of training content, however, but from efforts to structure the development environment. We determined that these efforts show promise and warrant the creation of another important approach to reuse: "structural" reuse.

Structural reuse occurs when an organization adopts something as simple as templates or style sheets or as complex as a complete content management environment (an LCMS, for example). Structural reuse also includes sharing processes (including Web services) that streamline procedures. In the most general terms, structural reuse is any computer code designed to make the development environment more cost-effective.

Commercial-sector case studies have shown a high degree of cycle-time reduction and development savings from structural reuse within individual companies. For example, three organizations realized large savings in cost and development time by employing technologies to automate the development and delivery of content in multiple delivery formats (e.g., online courses, job aids, instructor guides, lesson plans, classroom visuals, tests, and handouts) using a large central repository and one-time development of content (Chapman, 2007).

Whether communities of organizations (e.g., in the DoD environment) can successfully structure their reuse environment to obtain similar savings within an RLO repository framework has yet to be determined. Doing so would require a high degree of up-front collaboration to agree on a common TD environment but could promise large savings in development costs and production-cycle times. The need for strategic planning and collaboration in order for even ad hoc reuse to be successful is further discussed in Chapter Five.

The Decision to Bypass an RLO-Based Reuse Approach Makes Sense for Many Training Development Organizations

For many organizations, the decision to bypass an RLO-based reuse strategy appears to make sense economically, at least in this early adopters phase of technological development. At this point, implementation of an RLO-based reuse initiative requires significant up-front investment and organizational change while knowing that, according to the experience of our respondents, any returns are years away and by no means guaranteed.

Part of the risk of undertaking the RLO-based strategy is the need for success with different elements of the change-management policies—i.e., employing new technologies, successfully collaborating with other organizations, changing instructional design strategies and business processes, dealing with copyright and security issues. A second part of the risk has to do with the strength of the demand for reuse within an organization. One large organization that has a wide range of training needs and training suborganizations (and is a pioneer in the idea of employing reuse to reduce development costs) found that even after careful consideration, it significantly overestimated the extent to which the content it produced could be cost-effectively reused internally. It found that its typical digital training material simply had too much context to meet the needs of others with similar training needs.

Another part of the risk has to do with the effect on current customers. Some organizations determined that employing a reuse strategy would require the unacceptable cost of compromising service to immediate customers. A few respondents reported that incorporating the needs of those outside the customer base often increased the cost of the design of e-Learning or the amount of time needed to develop the courseware. A representative from one organization reported that the attempt to implement a strategy of reuse resulted in a substantial increase in time to development. Another respondent acknowledged a small increase in the production cycle; however, because this organization operates within a market whose customers expect a short development time, even a small increase in time to development was unacceptable.

Some TD organizations simply judge that returns internal to their own organization would be too small to justify the up-front investment. For example, a reuse strategy may appear inappropriate if the set of capabilities for which an organization develops training appear to be too broad and few potential partner organizations are available. Alternatively, the original design characteristics of training content (e.g., the amount of context built into learning objects) may make available training material a poor candidate for reuse from a technical standpoint. Even when there appear to be a great deal of overlap and no obvious technical issues, actual returns from reuse might not be large enough.

For example, in designing training to reduce oil spills from ships, the Navy studied the extent to which training lessons might be reused across its fleet (Concurrent Technologies Corporation, 2003). After detailed study of 22 Navy ship classes using eight oil-carrying systems, the Navy found that only 22 percent of the training content could be shared across hulls. While the effort was successful, the amount of actual reuse turned out to be fairly modest for such an ideal candidate area, and "the return" was lower than expected after accounting for the up-front costs of analysis, planning, and coordination.

For some organizations, other approaches to reducing development cost may have greater returns and less risk than an RLO-based reuse approach offers. Market evidence suggests that the cost of producing IMI has been decreasing over time.[1] From the viewpoint of the DL market, new tools and approaches are emerging all the time. For example, a representative from one organization said that IMI reuse had become of secondary importance once virtual schoolhouse technologies had emerged within the organization. From a technical standpoint, rapid authoring tools (independent of tools that allow reuse) are making e-Learning less costly all the time. In general, to the extent that improvements in technologies and authoring tools reduce the cost of developing e-Learning from scratch faster than they reduce the cost of reusing content, the case for reuse becomes less compelling.

Another approach to addressing the potentially high costs of DL development is to focus on processes that keep costs low to begin with. For example, when an organization's e-Learning needs are relatively simple and are unique or short lived (as when content is changing rapidly), a strategy of keeping production costs low may make more sense than investing in reuse. One organization employed such a strategy. Starting from a PowerPoint foundation and using an effective instructional design, the organization employed easy-to-use authoring tools to add flash and video, interactivity with students, and simulations. The organization felt that the resulting content was compelling and was inexpensive to produce and maintain. Moreover, this strategy precluded investment in an LCMS or more difficult-to-use authoring tools to support the repurposing of existing content.

As yet another example of an alternative approach to cost reduction in e-Learning, one organization focused its cost-reducing efforts on the delivery, rather than the development, side of its training (including e-Learning). Under this scenario, the employment of cognitive task analysis and a highly contextual training design model made e-

---

[1] One study estimated that the average cost of custom content development in 2006 was half of what it cost in 1999. See Chapman, 2007, and Brandon Hall Research, 2006a.

Learning content a poor candidate for RLO-level reuse. Use of cognitive task analysis, in fact, typically increases the design and development time by approximately 20 percent. However, performance-based evidence suggests that student learning time can be substantially reduced when this design model, rather than other design models, is used, leading to substantial net savings (Clark et al., 2006).

Reuse prospects might change if industry-wide repositories containing large amounts of content were to emerge. However, there is evidence to suggest that much of the content would likely not be suitable for reuse because the demand for any particular learning content would be too small.

Consider the case of a large defense organization that we interviewed. Figure 3.2 shows the number of graduates in 64 high-priority DL courses for this organization. Note the following about this group of courses:

Figure 3.2
Example: Distribution of Number of Graduates for High-Priority e-Learning Courses Within a Large Defense Organization

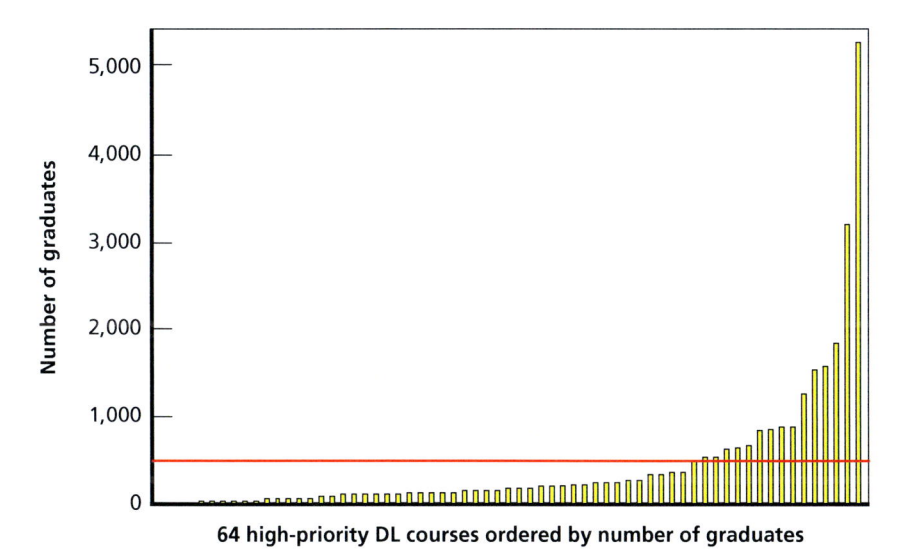

- Average graduation was relatively modest—only 422 students per course on average, with a median of 180.
- Only six courses had over 1,000 graduates.
- Twenty courses had fewer than 100 graduates.
- The potential for reuse outside the organization is seen as limited.

The average graduation across these courses suggests a small market for reuse. Some courses, such as those related to specific military tasks, have little application outside the military, and the potential for reuse in other military organizations is apparently small. The courses that do have civilian counterpart occupations in the public workforce could have potential application in the civilian sector; however, civilian access to the military repositories of the ADL-R will not be easy.

Experience with two existing large digital repositories suggests that only a few items in the DoD context are substantially reused (see Figure 3.3). In 2006, the Defense Audiovisual Information System (DAVIS) contained over 454,000 images and 13,000 film/video "active titles" that included content back to the 1950s. DAVIS acts primarily as a library/archive for use by DoD and public users. The Defense Instructional Technology Information System (DITIS) contains computer-based training and IMI content dating back to the mid-1980s. There were roughly 4,500 titles in DITIS at the time of this study. Like DAVIS, DITIS contains content descriptions and either allows the user to order the content directly or provides contact information for the content's owner so that the user can ask for access permission. Within the military service, it is required that new training content be registered with DAVIS/DITIS per DoD Instruction 1322.20 (DoD, 1991).[2]

---

[2] For example, U.S. Army TRADOC Pamphlet 350-70-2, *Training Multimedia Courseware Development Guide* (Headquarters, TRADOC, 2003) requires that developers thinking about developing new training content first check DAVIS/DITIS. In addition, the pamphlet directs that once content is developed, specific information on IMI that identifies the IMI program and describes the program's software and hardware is to be submitted for DITIS.

Economic Incentives 31

**Figure 3.3
Example: Orders from DAVIS/DITIS Digital Repositories Suggest That Only a Few Repository Items Get Substantial Reuse**

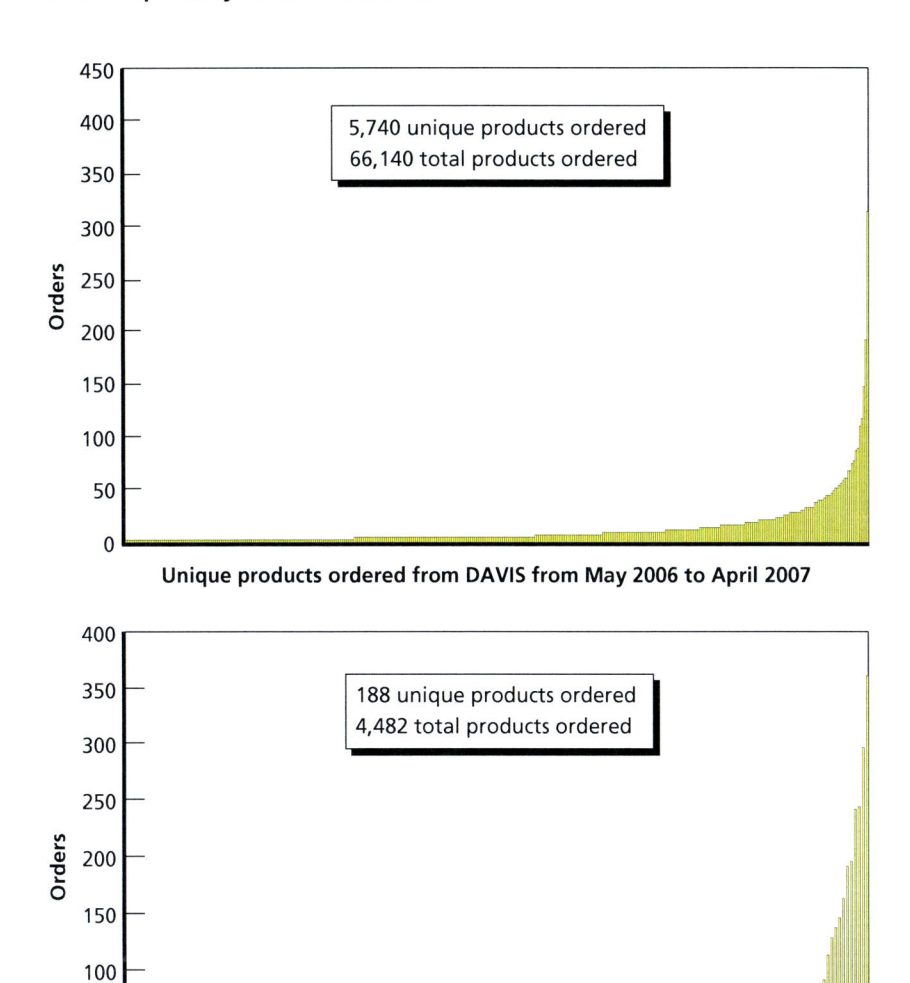

5,740 unique products ordered
66,140 total products ordered

Unique products ordered from DAVIS from May 2006 to April 2007

188 unique products ordered
4,482 total products ordered

Unique products ordered from DITIS from May 2006 to April 2007

RAND *MG732-3.3*

As Figure 3.3 shows, only a few of the items are frequently requested. The items ordered from DAVIS are more like "assets" for reuse in other contexts. The items in DITIS, which are complete courses, are presumably being ordered for reuse as a complete entity (top-down reuse) or, possibly, for concept reuse, but there is no way to gauge this split. Unlike other databases, DITIS courses are identified to the Defense Manpower Data Center for inclusion in the DoD database of formal courses.

Thus, we contend that much of the material that will soon be in the external repositories of the ADL-R (and perhaps other large repositories, as well) will see little reuse. This outcome suggests that developing organizations can potentially reduce costs (and thereby increase their ROI) by selectively designing their training for reuse: They would estimate the "reuse potential" of training before design begins and then invest in the extra cost of "designing for reuse" only in cases for which that potential is "high." Such an approach could reduce the cost of development without affecting global reuse in large repositories. Interviews with existing organizations suggest that the savings could be significant. In fact, overestimating the amount of material that had to be designed for reuse was identified by at least one organization as the major reason for abandoning its reuse strategy.

Some Digital Markets/Repositories Have Succeeded, Others Have Failed

Because the size of existing repositories for education and training objects is limited, we looked at established reuse markets outside e-Learning to identify insights on successful reuse strategies. We found that markets and repositories for digital content have had mixed success with reuse. Table 3.1 shows our findings.

On one end of the spectrum are markets and repositories (e.g., the multibillion-dollar commercial Web-based visual and audio programming industry) that successfully provide a high-quality product to a large consumer base:

Table 3.1
Successes and Failures of a Range of Digital Markets/Repositories

Name	Content	Type of Market	Relative Success	Key Characteristics
Asset markets	Visual, audio programming	Web-based commercial	High, multibillion-dollar market	High quality, niche markets, market analysis
COTS courses	Leadership, IT, other courses	Commercial	High, multibillion-dollar market	Large markets, high quality, market analysis
Online teaching— MERLOT[a]	Courses, learning materials	Open source and fee based	High	High quality
Global Health Network University	PowerPoint lectures	Open source	High	Niche market, great need, SMEs, easy to customize
Army 3D model repository	3D models	Free, not well known	Low	Difficult access, low quality, few entries
DAVIS/DITIS	IMI, videos, images	Free, not well known	Low	Difficult access, hard to customize, variable quality, not all entries

[a] Multimedia Educational Resource for Learning and Online Teaching.

- *Asset markets.* The Internet offers access to millions of both public-domain and privately licensed digital assets, including photographs, music, sound effects, and 3D models. The stock-photography business model is reportedly over 80 years old, and its current realization on the Internet is approaching $1 billion in annual revenues.
- *COTS courses.* Commercial companies offer a variety of Web-based courses for a fee, especially those covering the areas in high demand across organizations (e.g., preventing sexual harassment in the workplace, leadership development, regulatory compliance, and safety). There are also Web-based courses in specific mar-

kets with an emphasis on IT knowledge/skills and use of software applications.[3]
- *Online teaching support sites.* Organizations such as MERLOT[4] provide open-access, Web-based referatories for Web-based learning materials of interest to higher-education students and faculty. In the case of MERLOT, the site offers some content for free and some for a fee. Anyone can contribute content, but peer review for quality assurance is often part of the process when content is made public. There is also a public mechanism for providing comments on content.
- *Global Health Network University.* Also known as Supercourse,[5] this repository offers about 3,000 lectures on public health (as of December 2006) from recognized experts in their field, all free of charge. Despite the small size of this repository, managers report over 40,000 users from 171 countries, most of which were experiencing public health crises.

On the other end of the spectrum are repositories that enjoyed much less success. However, since failures, especially market failures, are more difficult to find because they tend to be small and disappear once they fail, we are able to describe only the DoD's digital repositories for training content:

- *Army 3D model repository.* Sponsored by the Army's Program Executive Office for Simulation, Training and Instrumentation (PEO STRI) Targets Management Office and the Army's Research, Development and Engineering Command (RDECOM) Virtual Targets Center, this repository was designed to collect and share

---

[3] See, for example, the offerings at TrainingTools.com (undated), Course Technology (undated), and BestWebTraining.com (2008).

[4] Originally developed by the California State University Center for Distributed Learning, MERLOT was based on an "educational object economy" concept developed in work done by Dr. James Spohre and funded by the National Science Foundation. See Multimedia Educational Resource for Learning and Online Teaching, 2008.

[5] Supercourse was developed by Ron LaPorte at the University of Pittsburgh. See Supercourse: Epidemiology, the Internet and Global Health, 2009.

Army-developed 3D models.[6] However, because contribution to this repository is not required, few additions have been made, and this digital source has reportedly received little use. Also, during a review of its content in October 2007, the models were judged to be of low quality compared with what was available on commercial markets.[7]

- *DAVIS/DITIS.* TD organizations are required by DoD Instruction 1322.20 (DoD, 1991) to use DAVIS/DITIS, a repository system that, as of May 2007, had 20,000 registered users and content that included 454,000 digital images and 13,000 videos, as well as 4,500 CD-ROMs containing IMI material. We classify this system as an example of less than full success. The number of searches has been estimated at about 18,000 a month, resulting in 2,000 product orders per month, with an average of two titles per order. Actual usage of the titles is unknown, but anecdotal reports from interviews with Army schools suggest that there is little value in searching DAVIS/DITIS. Moreover, the value of the repository and its content are questionable. Visual information goes back to the 1950s, and CD-ROMs go back to the 1980s. Importantly, the system is not nearly as complete as intended, because despite the DoD instruction, organizations are not strictly required to contribute in practice (and many choose not to). In consequence, searches are less likely to find what is being sought than they would be were the system kept current. Furthermore, access is limited by the fact that the remote search capability allows key word search only and that content cannot be viewed before ordering. Finally, improvements to the system and its management are difficult to make because no metrics describing successes and failures are kept.

---

[6] The U.S. Army Model Exchange Web site is available only to account holders.

[7] Detailed 3D models are commercially available online at low cost from a number of sources—e.g., Digital Dream Designs (undated), The 3D Studio (undated), TurboSquid (2009), and Google SketchUp (2009).

By looking at existing repositories, we were able to draw a number of lessons about what customers demand. In general, markets succeed when they are user driven and well supported. More specifically:

- *Quality and service matter.* If a repository's content is not high quality, the venture is unlikely to succeed. The success of a commercial asset market and the willingness of its customers to pay depend on quality, and lack of quality appears to be part of the explanation for repositories that do not succeed. Thus, we conclude that the value of low-cost, high-quality content trumps that of free, low-quality content.
- *Demand is the most important determinant for success.* The experience of the Supercourse initiative shows that success requires content for which there is high (or even urgent) demand. For example, COTS applications in leadership and IT are successful because they apply to a wide range of users/environments. High demand in turn depends on the use of widely accepted paradigms and formulations, such as those used by MERLOT and Supercourse. High-demand material will undoubtedly include some RLOs.
- *Open-source and nonprofit repositories can work.* The MERLOT and Supercourse examples show that repositories with potential "public good" issues can be successful if carefully designed and managed. For example, MERLOT shows that nominally nonmonetary incentive structures can be used to obtain contributions (e.g., MERLOT offers a venue in which academics can publish) and that the quality of contributions can be managed within a not-for-profit environment (e.g., through peer review).[8]
- *Market analyses can align supply and customer demand.* Those on the supply side must continually analyze the changing nature of demand in order to meet customers' changing needs. Market analyses include measuring the degree of success, soliciting customer feedback, and committing to continual change. Reposito-

---

[8] Other examples also exist, including the Institute for the Study of Knowledge Management in Education's (ISKME's) Open Educational Resources (OER) Commons (OER Commons, 2009) and the National Registry of Online Courses (NROC) Commons (NROC Commons, undated).

ries that lack market analyses have a history of under-utilization and under-contribution (Neven and Duval, 2002).
- *Transaction costs can heavily impact the degree of success.* The "transaction" costs of reuse derive from how difficult it is to access and customize digital material. DAVIS and DITIS were not designed to support reuse and appear to provide limited value to those seeking interactive training content, at least partly because of the relatively high transaction costs involved. Access is difficult because the only way to search is by key word, and content is neither searchable nor indexed to provide more-specific information. Thus, customers must go through the entire process of ordering and waiting for material before determining whether it fits their needs. Moreover, customers who have received material cannot easily customize it, because the underlying program or digital assets are not included.

History of Attempts to Reuse Software Provides Insight on Training Content Reuse

The commercial software industry provides an example of a much larger reuse market. Thirty years of attempting to package solutions to ongoing problems at all levels, plus the fact that e-Learning is, in fact, largely software, make software an important source of lessons for reuse within e-Learning. Reusable software content can take many forms, including complete applications, subroutines, functions, macros, libraries, objects, and design patterns. Although there are multiple benefits to creating such entities, a key benefit is the ability to reutilize their capabilities in the future without having to redo the analysis, design, and programming that went into their creation.

The software market is larger than the e-Learning market, because software spans all commercial, industrial, governmental, and personal sectors and functions, including (but not limited to) e-Learning. In addition, it may be easier for some types of software to be made widely reusable, since some may have wider appeal (e.g., it may be easier for diverse communities of users to agree on the requirements of a payroll system or a cosine function than it is for education districts to agree on the content and pedagogy of a math course).

Although e-Learning is, for the most part, software, the production of software is different enough from the production of e-Learning to affect the amount of reuse that occurs. One difference is the average difficulty and expense of production. Through the use of authoring tools, quality educational material can be produced competently by a much larger number of practitioners than can software. If quasi-digital educational material (which makes little use of the capabilities of the digital medium) is included, an even greater number of people can produce e-Learning material. Considerably more training and experience are typically required to produce a successful software product. One reason is that software tends to be more intricate and less forgiving of errors than is e-Learning, for which more-informal design and implementation can often produce acceptable results. To the extent that e-Learning relies on textual or graphic content, it can be produced effectively by most competent teachers in a given subject area, whereas software requires not only subject matter expertise, but also highly specialized programming skills. Furthermore, e-Learning material need only be applicable across a relatively limited range of subject areas and contexts, whereas software may have a far more general function (such as performing optimization or data mining) that may be invoked across a much wider range of purposes and contexts. Software must therefore be more robust and general purpose than e-Learning. As a result, software tends to be more expensive on average to produce than is e-Learning.

However, the market for educational material is better defined and understood than is the market for software. Texts, classes, courses, and seminars already exist for a huge range of subjects and have for a long time, which means these have relatively well defined and mature markets. In contrast, software, even though it is evolving and expanding rapidly, has existed for only 50 years. There is intense competition for novel functionality and new products to carve out new software market niches—along with frequent failures in trying to do so. This volatility and evolution of the software market have made the production of reusable software a highly speculative undertaking that involves considerable intellectual and financial risk.

The implications of these differences go in opposite directions—one suggesting that software reuse would represent an upper bound on e-Learning reuse (because it is more expensive to develop software than to develop e-Learning), and the other implying that software reuse may represent a lower bound on e-Learning reuse (because of the greater risk involved in attempting to develop reusable software).

Experience with Software Reuse Suggests That Success Will Occur Only in Selected Cases and That They Will Be Difficult to Identify Beforehand

The notable successes in reusing software have often required clearly defined and robust communities of potential users (Schmidt, 2006). A community of users sharing a given programming language, computer platform, operating system, or application development environment can often provide a reasonably robust market for reusable software. For example, programming-language-specific modules (such as Java and C++ class libraries, Enterprise JavaBeans, and Visual Basic code snippets) have achieved significant levels of reuse, although their quality varies greatly, and reuse of these resources is often limited to a specific context, platform, or environment. The open-source software movement has similarly produced large amounts of code that have seen widespread reuse, particularly in the Linux environment; but here, too, quality tends to vary.

The most spectacular reuse success stories in software involve complete application programs (e.g., word processors, spreadsheets, database management systems) and operating systems, which are reused extensively. However, such large-grained cases are not normally thought of as reuse—in e-Learning, these would correspond to the reuse of entire courses or e-Learning environments. Furthermore, there have been notable failures even among complete applications. Enterprise Resource Planning (ERP), for example, has had very mixed success, because any given ERP system has to be customized to the specific needs of the company attempting to use it.[9]

In addition, despite some communities' positive experiences with fostering reuse, others have been less successful. For example, the Ada and Corbus communities have had limited success in this regard, and the Defense Advanced Research Project Agency's Megaprogramming effort (in the early 1990s) was a resounding failure (Wiederhold, Wegner, and Ceri, 1992; Boehm and Scherlis, 1992). The reuse of COTS and government off-the-shelf products within the U.S. military (and the government in general) has also had mixed results, in this case greatly reducing initial cost, but often leading to problems stemming from poor fit to the adopting organization's needs, the attendant need to perform (and maintain) costly customizations over time, and undue reliance on commercial software developers and vendors, whose priorities are driven by the commercial market rather than government needs.

With the exception of complete application programs and operating systems, finding positive ROI has been the exception in the software development industry and has proven difficult to predict (Brynjolfsson, and Hitt, 2003; Brynjolfsson, 1993; Koenig, 1993; Roach, 1991, Pfleeger, 1988, Pfleeger and Cline, 1985). One of the most significant stumbling blocks to making software reusable is the task of balancing its functional generality against users' need to customize its behavior to their specific requirements.

The more general purpose a given software component is, the wider its potential market for reuse should be, but the effective utilization of such functional generality often requires users (or user organizations) to customize the component, selecting and adapting its wide range of functional capabilities to their specific needs. In some cases,

---

[9] ERP systems consist of complex software applications used by large enterprises to manage inventory and integrate business processes across multiple divisions and organizational boundaries. For further information on ERP systems, see "E-Business Insight—ERP, CRM and Supply Chain Management," 2005, the home page of a knowledge base and news aggregator on the implementation and integration of business software. For a discussion of the causes of ERP system failure, see "Causes of ERP Failures," accessible via the "ERP" link on that home page.

little or no such customization is necessary, such as when reusing simple mathematical or scientific functions—e.g., sine or square root in the widely reused NAG (Numerical Algorithms Group) Fortran scientific subroutine package.[10]

But in most cases, some customization is required; and in many cases (such as ERP), an extreme demand for customization may undermine reusability. Unfortunately, no simple rule has emerged to indicate which kinds of functions can be made general enough to have a wide reuse market while requiring minimal customization and enabling developers to exploit their market effectively.

There appear to be many possible "sweet spots" that combine different degrees of generality and customizability. Sweet spots are determined by numerous attributes of the content, context, granularity, and functionality of the software in question, and it has proven difficult to predict where they will occur. Yet in order to be reusable, software must accurately characterize and explain its chosen sweet spot—i.e., the point that appropriately balances generalization and customization. Furthermore, it must document its intended context of use and the semantics of both its processing algorithms and its interfaces, using formalized ontologies to enable users and other software to judge how suitable it may be for reuse.

In e-Learning, the need for customization may be even greater than it is in software in general. This is because e-Learning embeds not just functionality, but also terminology, semantics, world view, pedagogy, subject matter, disciplinary context, and numerous other aspects that may be crucial to the effectiveness of training and learning. Potential users of e-Learning software have expressed an especially great need to customize its "look and feel" and behavior. Thus, the choices made by an originating training organization to put its "brand" on training content may well be highly objectionable to potential reusers of that content, even if the branding does not substantially alter training. This tendency has no direct analogy in the wider, software reuse market.

Another obstacle to reusing software has been the multiple ways in which content can be organized, or "factored," in achieving a given

---

[10] For a description of the NAG Fortran Library, see Numerical Algorithms Group, 2008.

goal (Page-Jones, 1980; Yourdon and Constantine, 1979). For example, software can be designed by dividing and packaging its functions according to their order of execution (phases), type of data, type of operation, "tier," etc. Re-factoring software (i.e., rethinking the way it is organized and packaged) is very expensive. In addition, because factorization does not easily capture "cross-cutting" concerns (such as interactivity, customizability, reliability, persistence of results, and extensibility), approaches have had to be developed to represent such concerns "orthogonally" to the primary factorization of a product (e.g., aspect-oriented programming [Kiczales et al., 1997]).

For e-Learning, these problems are likely to be even greater than they are for software in general, because the additional social, political, and cultural factors associated with e-Learning material suggest that factorization will be even more crucial to its success. World view, explanatory paradigm, context of application, degree of detail, terminology, pedagogic approach, and intended audience are but a few examples of factors likely to complicate widespread reuse in e-Learning: Each potential context in which a given e-Learning product is to be used may require customization to adhere to a different set of such factors.

In software, it is expensive to engineer for reuse (i.e., to design to meet the needs of a wide range of unknown users) and to integrate reusable components into a target system, but failing to do either of these makes reuse unlikely to succeed.[11] As an example of the need for integration in the e-Learning context, consider the fact that many organizations in the military need to train individuals in self-protection, making such training a candidate for reuse across many different TD organizations. However, some organizations may need to integrate such training into an existing course as a two-hour block of instruction, whereas others might need to make it a full day or more. These different intended uses require that a given module be customized and integrated into quite different contexts, which is no easy task.

---

[11] Previous RAND work on this issue includes an analysis of the problems involved in component-based modeling. See Appendix B, "The Elusive Nature of Components," pp. 86–91, in Davis and Anderson, 2003.

Similarly, some schools might teach a task as a stand-alone lesson or course, whereas other schools may embed the training for a particular task in a large lesson plan that instructs on multiple tasks at the same time—and that tests multiple tasks at once in the form of practical exercises. Integration may be less difficult if reuse focuses on less-detailed knowledge within a task (e.g., defining self-protection and general approaches to it), but then the training is likely inexpensive to reproduce from scratch, making reuse less compelling.

Another aspect of integration is the process of adapting and integrating a given e-Learning module into the learning system in which it is placed. Software intended for use only by its developers need not be nearly as robust as software intended for reuse by others. A well-known rule of thumb in the software world states that turning a program with a given functionality into a reusable component that is thoroughly generalized, tested against potential misuses, and thoroughly documented takes roughly 10 times as much effort as producing the original program (Brooks, 1979).[12]

Many of the points about software can be illustrated using an example of software reuse in the open-source arena. Successful reuse has occurred with significant amounts of open source Java and C++. This success might suggest that e-Learning software could similarly be reused within the context of large public e-Learning repositories. However, the overall experience with software reuse in open-source contexts has, in fact, been mixed, with both many failed efforts and many successful ones (Glass, 2001). The problem arises because open-source artifacts are not typically designed for reuse; instead, they are simply published once they become (often barely) operational, in the hope that someone else can use them or improve them. This results in extreme variance in the quality of these artifacts, which, in turn, mixes successful reuse with failures. Further, when success does occur, it often depends on the considerable skill of programmers, who must customize and integrate the material to fit a new context. When suc-

---

[12] For more-recent and more-quantitative comparisons of the effort required to produce software of various kinds, see Boehm et al., 2005; Succi and Baruchelli, 1997.

cess occurs and the skill is marginal, the effort produces new code that is even more poorly suited for further reuse.

In the e-Learning context, a similar variance in quality is also likely to exist in large public repositories containing digital training and educational material, and a similar need for integration and customization will exist. However, as already pointed out, the skill required to customize and integrate may be higher in the software world than is typically available or warranted in the e-Learning context. Thus, we argue that the software experience of reuse in open-source contexts illustrates our point that reuse in the e-Learning context will be successful only in select cases and, perhaps, not to the same extent as in software.

History of Reuse in Materiel Design Provides Insight on Training Content Reuse

Reuse in the materiel design area provides another potential source of lessons learned for e-Learning. RAND research by Newsome, Lewis, and Held (2007) documents a long history of attempts to use common components in the development of multiple products. Industries as diverse as the automobile, aircraft, military vehicle, and weapons system industries have been successful (and sometimes unsuccessful) in their efforts to realize the benefits of commonality.

The goals of materiel reuse are similar to the goals of e-Learning reuse—to decrease development, maintenance, and training costs while minimizing any negative effects on operational capability. Success in reusing designs in materiel development is typically expressed in percentage terms and referred to as "commonality levels" reached.

Reuse in materiel development can occur at various levels, from the systems-of-system level down to the part level. As an example, consider the M4, the current Army and Marine Corps rifle issued to most ground forces. An end-item such as the M4 is made up of components (e.g., the night-vision sight), subcomponents (e.g., the trigger assembly), and parts (e.g., a spring).

The levels of materiel redesign can correspondingly be expressed in e-Learning in terms of assets, educational learning objects, lessons, courses, etc. Thus, continuing our M4 example, we find the following:

- *System.* At the system level, materiel development might focus on an M4 rifle with a night-vision scope. A system-level e-Learning element might be a course on changing the tires of a Humvee (High Mobility Multipurpose Wheeled Vehicle).
- *Component.* The night-vision scope for the M4 rifle is an example of a component—it can stand alone as an entity but functions within the context of a system. In e-Learning, a component-level element would be an educational learning object, which can stand alone to teach a single skill (e.g., tightening lug nuts on a wheel), but functions within the context of a course on tire changing.
- *Subcomponent.* The rifle's trigger assembly is an example of a subcomponent—it comprises multiple parts and has no function outside the context of the next-higher level, the components. A parallel in e-Learning would be a reusable object, such as a small animation or simulation that demonstrates how torque increases as the level arm lengthens.
- *Part.* A spring provides an example of a part—an individual piece or set of pieces that can only be ordered as a single item and that has no separate use outside the context of higher-level components. In e-Learning, a corresponding element would be an asset, such as a photograph or graphic that appears on a Web page.

Like digital training content, materiel development has some clear successes from seeking commonality, or reuse, in materiel development. For example:

- The Joint Strike Fighter has three variants, all made possible via a component-based strategy for engines, avionics suites, and portions of the fuselage:
  – The U.S. Air Force's F-35A, CTOL (conventional takeoff and landing) variant

- The U.S. Navy's F-35C, CV (carrier-based) variant
- The U.S. Marine Corps's F-35B, STOVL (short-takeoff and vertical-landing) variant

- The U.S. Army's Stryker Vehicle, made by MOWAG,[13] is based on the Piranha III chassis and offers options for power plant/transmission to enable commonality, as well as 10 variants for different Army uses.

But the materiel development world also has many failures and compromises of capability, illustrating the potential pitfalls of reuse. The several different ways in which there have historically been poor trade-offs of capabilities and commonality include

- *commonality by fiat,* through which a decisionmaker dictates that a certain level of commonality will be used. One such case that led to significant loss of money and product quality was the 1960s development of an aircraft that became the General Dynamics F-111. The design process was charged with developing a single aircraft to serve U.S. Navy and U.S. Air Force needs at the time for a medium-range strategic bomber, a reconnaissance aircraft, and a tactical strike aircraft that could take off from both land and aircraft carriers. This requirement to cover too broad a set of design specifications led to the Navy's withdrawal from the program, the production of significantly fewer aircraft, very high costs, and remaining operational trade-offs in the performance of the final aircraft.
- *commonality mediocrity,* which is driven by designers striving for too much commonality. A compelling example is the development of the U.S. Army's World War II medium tank platform. This platform supported a very wide family of armored vehicles, but the tank variant turned out to be too small and underpowered against heavier foreign tanks.

---

[13] MOWAG, a Swiss company, is now owned by General Dynamics and is part of its Combat Systems Group. See MOWAG, 2009.

- *building in excess capability*, which is an attempt to leverage a common component across products that can lead to monetary loss in manufacturing. A U.S. auto manufacturer, for example, produced a common wiring harness that was to be used across a variety of car models, both high and low end. Building this unused capability in the low-end models proved not to be cost-effective, so the company went back to separate, tailored wire harnesses.

The U.S. Army is contemplating a decision process for determining where there can be cost savings without inappropriately high operational losses in the design of new materiel. The process requires a series of top-down specification and planning steps that take planners through a "model plan," a "differentiation plan," a "commonality plan," and, finally, a "base model plan." Although unproven, the model is intended to prevent the traditional pitfalls of commonality decisions listed above and lead to more cost-effective materiel while minimizing operational sacrifices.

Just as an elaborated decision process and specific decision support tools can improve the effectiveness of decisions about acquisitions of Army materiel using "common" components, so too can they help with decisions about when to invest in reuse of "common" digital learning objects. The higher the probability of reuse, the more resources that need to be invested in developing the learning object so that both the immediate need and the possible needs of future reusers of the object are met. The criteria for identifying learning objects predicted to have a high probability of reuse include

- the level of pervasiveness across the Army and Joint Forces of the materiel or processes being trained
- the expected "life" of the content—short-lived course content has a lower chance of reuse.

Recommendation: Make ROI for Reuse an ADL Focus Area

The success of the emerging learning object economy may critically depend on the extent to which TD organizations are convinced of its value, as well as the degree to which early adopters of reuse are able to realize and report positive returns.[14] As previously argued, the perceived value of reuse is truly a key enabler, but it may need support in the early stages of the life cycles of reuse-supporting technologies. Our results show that a strong perception of the value of reuse has emerged from only a few of the early adopters of an RLO-based reuse strategy. Further, outcomes for reuse in contexts outside e-Learning suggest that successful applications of e-Learning (especially RLO) reuse are likely to be in specific areas rather than broad based.

Thus, to foster the success of the emerging learning object economy, we recommend that ADL make the ROI from reuse a specific area of near-term focus. ADL might foster reuse success and positive perceptions in three ways:

To build the economic case for reuse, ADL should broaden the definition and gather documentation from payoffs based on that wider view. Various repositories, including the ADL-R, will include metrics that seek to measure the frequency with which RLOs (or digital assets) are drawn for potential reuse. While measuring this reuse will be important, the ADL should also take steps to ensure that all five of the approaches to reuse identified in our study are included in the definition, and should seek to document payoffs in each area.

Measurement efforts should thus go beyond RLO reuse from repositories to include results from top-down and bottom-up approaches, many of which do not involve repositories. This information might be gathered using ongoing literature searches and select surveys and case studies.

For example, ADL should try to document ROI from significant efforts to reuse e-Learning material in technical documentation,

---

[14] Without the perception of high payoff, enabling behaviors, such as the formation of communities of interest that foster reuse within specific subject areas, are much less likely to emerge.

in performance support systems, and in other "use cases." Savings on course maintenance within organizations that are attributable to an RLO approach to design should also be documented. Savings derived from internal asset repositories within organizations might also be highlighted; for example, one organization in our survey reported doubling its profit from the strategic use of asset repositories.

ADL should also attempt to document ROI for efforts in which organizations collaborate to create courses that serve multiple audiences. For example, the Army is creating professional military education courses that serve across training organizations,[15] and the Department of Veterans Affairs (VA) has coordinated collaborative course designs across the services.[16]

In broadening the definition, ADL should also recognize, support, and attempt to measure the extent of another of the identified approaches to reuse, that of concept reuse. As stated previously, this approach entails the reuse of ideas (or structure, or format) as opposed to digital content. The advantage of concept reuse is savings on design costs, which are often a significant portion of overall development costs. Further, concept reuse can be implemented in the absence of great advances in technology.

As previously suggested, a key aspect of successful concept reuse is to devise and support ways of helping large referatories, such as the ADL-R, to provide faster access to e-Learning content. Currently, the ADL-R provides rapid access only to the metadata for the content. One way to provide faster access is to store actual content on the ADL-R, which may be possible in some instances. For cases in which content

---

[15] For example, for its training of non-commissioned officers (NCOs) in all occupations across 17 proponent schools, the Army has created a common-core phase in what it calls its "basic course" to provide the same instruction to all NCOs. The Armor School and the Infantry School are collaborating to produce a single "maneuver advanced NCO course" to replace two separate courses that exist today.

[16] The Veterans Health Administration Employee Education System has launched a collaborative effort among a group of federal agencies to develop reusable healthcare training content that has applicability across multiple agencies. While the up-front collaboration is an additional cost, substantial net savings can be achieved when multiple agencies pool resources to produce courses that fit a wide range of needs. For a further description of this effort, see Twitchell and Bodrero, 2006.

owners do not want to store actual content on the ADL-R, indexes of the content could be stored using commercially available content analysis software.[17] Aided by commercially available search engines, these indexes could provide users with a much fuller description of the content within a matter of minutes. An example would be to use the search capabilities of a company such as Google's commercial business services to index, cache, and provide access to all ADL-R content.[18]

ADL should also seek to document success stories in the area of structural reuse. As previously stated, structural reuse has already been documented within commercial organizations (Chapman, 2007). ADL would be interested in supporting and documenting reuse efforts from communities of potential reusers (i.e., efforts involving multiple organizations that attempt to structure their development environment to promote reuse).

ADL can support those considering a reuse strategy by helping to predict what materials are most likely to be reused—that is, where the "sweet spots" will be. Based on our findings about reuse in other domains, we conclude that the factors, or criteria, that lead to a higher probability of success include

- a big potential market for content in future-use cases
- feasibility within and among organizations (i.e., organizational readiness)
- no unresolved factorization issues
- an acceptable balance between generalizability and the need to customize
- high quality and low transaction costs.

Materials that appear to have a high reuse potential might justify a greater effort from the developer to design for reuse. For example, when designing for reuse, developing organizations might contact

---

[17] When indexing software is applied to Web searches by companies such as Google, it is called a Web crawler, Web spider, or Web robot.

[18] Google offers a variety of search capabilities to organizations at its Google Enterprise (2009).

organizations that could potentially benefit from reuse to seek input on design needs or possibly to collaborate on joint production of a course. They could also plan for a larger effort in the design process to separate content from context, structure, etc.; make greater efforts in producing detailed metadata; or make a higher-profile effort to let people know that content is there. Conversely, for materials that appear not to have a high potential for widespread reuse, organizations might be advised to deemphasize reuse. For example, for training that is highly contextual, for which demand outside the original course is likely to be relatively small, or that is easy to reproduce from scratch, developers might seek to reduce design costs related to allowing more-general reuse. This might be manifest in smaller metadata efforts (especially if content can be indexed), more flexibility in the grain size of RLOs (e.g., allowing larger grain size where appropriate), or less testing for SCORM compatibility with other LMSs.

ADL might make ROI a focus by sponsoring research on outcomes of reuse initiatives. One way this might be done is through case studies of particular efforts that look at ROI, incentives, and other implementation issues related to reuse. For example, ongoing efforts in the military acquisition area, the medical area, or multiple other areas may lead to opportunities for research that

- measures "return on investment" in promising areas
- generates guidelines on how to reduce the up-front costs of participating in reuse (e.g., where large grain size is acceptable)
- evolves ROI estimation and evaluation methods and metrics
- develops a decision process and decision support tools for allocating resources to develop "high reuse" content, similar to those suggested for use in Army materiel acquisition.

If possible, ADL should also look for an opportunity to evaluate a true market in reusable training objects that includes reimbursement to publishers whose objects are sold. While such a market is not consistent with current DoD policy, the prospects and problems of the concept might be studied by ADL in other areas, such as academia. The example of the open-source market, including fee-based options for acquiring learning objects (such as implemented within the MERLOT referatory for higher education, online learning materials) provides a precedent for potential development of a market for RLOs.

Another way to promote research on the ROI from reuse is to develop additional survey data and metrics within the ADL-R. Repository owners have two obvious points at which they can collect data: when producers submit or supply content to the repository, and when users search for content. In either case, pop-up surveys could be used to ask short questions. In the case of producers, surveys could ask whether they designed their product for reuse and whether they made use of reused materials in their development. In the case of users, surveys could ask what they are looking for in coming to the repository. Post-search surveys (as conducted by many online companies) might ask users if they were satisfied with their searches and, if not, whether the dissatisfaction stemmed from the content itself or the transaction costs. Collecting data of this sort would put ADL in the role of market analyst for reuse initiatives. As trends emerge, repository design can be adapted to provide more of what customers need. In this way, ADL would be supporting the spiral development of a reuse market.

CHAPTER FOUR
Disincentives to Sharing

This chapter considers potential disincentives to reuse that arise within larger TD organizations or industries. Disincentives can become obstacles to reuse, whether they arise from stakeholder reluctance to share learning objects or to reuse content created by others. Our analysis drew both from the study's interviews with TD organizations and from the literature of KM, which has a longer history of addressing obstacles to the production and sharing of intellectual assets.

Our overall conclusion about disincentives is that they are currently of secondary importance to stakeholders as obstacles to reuse in the e-Learning context but could threaten future successes. We recommend that ADL pursue options to foster the establishment of positive incentive mechanisms for implementing a reuse strategy within TD organizations.

Various Stakeholders Participate in a Strategy of e-Learning Reuse

By *stakeholders*, we mean subgroups associated with TD organizations that are involved with the TD process. The stakeholders in our survey and analysis included TD headquarters organizations, TD suborganizations, custom content developers, and employee groups within those organizations (e.g., program managers, instructional designers, computer experts, and SMEs).[1] Figure 4.1 portrays the typical relationship of these stakeholders at the time of our study. The figure shows one large TD organization with several suborganizations. These suborganizations may develop content in house or contract it out to one or more custom content development organizations. A custom content developer might work independently or might have its employees collaborate (and potentially even co-locate) with the staff of the TD organization.

Training Development Organizations Considered Disincentives to Be Secondary Obstacles to a Successful Reuse Strategy

In our telephone interviews with TD organizations, we asked whether the issue of "incentives among stakeholders" represented an obstacle to their reuse strategy, which stakeholders showed reluctance to participate, and what respondents thought caused that reluctance. We also asked whether the respondents were aware of any measures in place to address the issue of incentives. In some interviews, representatives from each stakeholder group were present to speak for their group; in other situations, non-members of stakeholder groups represented the issues.

Many TD organizations in our survey noted some current disincentives to reuse; 55 percent said that some stakeholders had "less than full enthusiasm." However, these factors were typically not cited as critical in impeding the development of a reuse initiative. Only one TD organization in our sample cited disincentives as the "greatest obstacle" to reuse.

The most commonly cited disincentive, "Do not see significant benefits in reuse," appeared to be closely related to the ROI issue discussed above, and applied to both the production of content and the reuse of others' content. It appeared that many of the individual stake-

---

[1] As explained earlier, we use the *TD organization* or *respondent* to mean all or the majority of participants in an interview that are associated with a TD organization. Typically, interview *participants* also belong to a particular stakeholder group.

Figure 4.1
Stakeholders Associated with TD Organizations and Their Potential Relationship

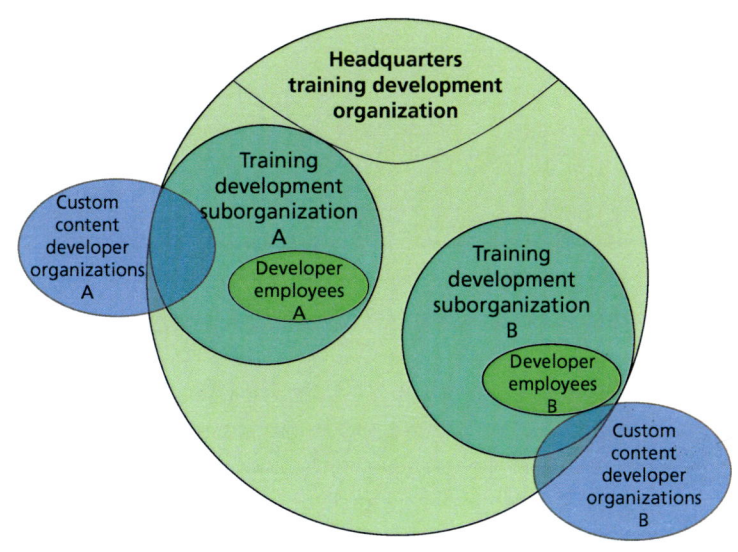

holders held the same views as larger organizations regarding the low potential for ROI from reuse.

Another disincentive to designing for reuse was that it would involve significant work (e.g., in producing metadata) that would not be compensated and that would potentially be at the expense of a current customer (e.g., if an immediate customer had to wait longer for the product or received a product that, because it had been designed for general reuse, was less valuable than it would otherwise have been[2]). This problem is related to the first disincentive because costs do not have to be high to result in an unacceptable cost/benefit ratio when benefits are believed to be close to zero.

In the several cases in which respondents saw some benefits to reuse, they usually noted the free-rider problem involved—i.e., that

---

[2] For example, a customer of one TD organization was dissatisfied with the small grain size of the reusable piece of software he received because he preferred more training between tests and fewer required "clicks" for getting started with the training.

while they, as developers of e-Learning, would accrue all the cost of designing for reuse, others, who had paid none of the original cost, would capture a good deal of the benefits. Thus, in an open repository, a valuable piece of reusable e-Learning becomes what is termed a *public good* in economics.[3] It is important to note that according to basic economic theory, a public good will tend to be under-produced unless free riders can be made to pay, higher levels of production can be mandated, or alternative incentives can be brought into play.[4] This chapter deals with the third of these options.

Two organizations told us that because they considered training to be part of their competitive advantage, they were unlikely to collaboratively produce common training or to contribute e-Learning products to industry-wide repositories even though companies in their industries had common training needs. These respondents felt it would take particularly high returns from reuse to provide a sufficient incentive for collaboration on training within their industries.[5]

A moderate number of organizations noted a disincentive issue involving custom content developers that would increase as the learning object economy matured and reuse became more common. These organizations noted that while custom content developers hired by TD organizations were currently cooperative and occasionally proactive with respect to reuse, they would lack sufficient incentives to comply with the "spirit of reuse"—i.e., to produce a sufficient amount of highly reusable content—if the learning object economy were to expand significantly.

An important part of developers' current business model is to resell material from their own proprietary repository to new custom-

---

[3] A public good is one for which consumption of the good by one individual does not reduce the amount of the good available for consumption by others—i.e., no one can be effectively excluded from using that good. For further explanation of public goods, see standard textbooks in microeconomic analysis (e.g., Varian, 1992).

[4] Note that if the cost of designing for reuse is reduced (through improvements in technology and the evolution of cost-effective processes), the free-rider problem will lessen.

[5] Proprietary training is not an insurmountable challenge. Later in this chapter, we cite an example of private organizations sharing proprietary information in order to accrue greater perceived gains through collaboration.

ers. One developer noted that his organization's profit margin doubled from its ability to reuse material (primarily digital assets) it had previously produced. If a successful learning object economy tended to shrink these margins, custom content developers would naturally look for ways to make the content they contributed to these repositories more difficult for others to reuse.

Disincentives May Become a Bigger Issue If Reuse in e-Learning Becomes More Prevalent

Despite the fact that disincentives turned out to be secondary-level obstacles in our study, we anticipate that if repositories become more prevalent and reuse becomes more common, disincentives to sharing content or reusing the content of others will become a more significant problem. Research in knowledge management systems (KMSs) provides the foundation for this expectation.

Training content is an intellectual asset. As such, understanding the processes underlying reuse of training content can be informed by the theoretical and empirical literature on KM. KM refers to the processes that organizations use to manage their intellectual assets. KMSs, which typically consist of computer-based systems (Alavi and Leidner, 2001), help people capture, develop, organize, and distribute knowledge and information. Research on KM has investigated factors that affect whether individuals or organizations will share knowledge or assets, as well as factors that influence whether individuals will use knowledge or assets that others have shared.

Knowledge Sharing

Organizations and individuals can be reluctant to share knowledge for various reasons. One reason is the effort required to codify knowledge to make it usable to others; another is concern about free riders, those who will use but not contribute to a shared resource. A third reason is the concern of individuals that they will lose status or credit for ideas and the concern of organizations that they will lose a competitive advantage. Fourth, individuals may feel that knowledge systems serve as a replacement for their work, leading them to feel that people are expendable or less valued as employees (Evangelou and Karacapilidis, 2005). Finally, roles and values also affect knowledge-sharing activities. For instance, lower-level employees may hesitate to offer knowledge out of the expectation that senior co-workers will not perceive the contribution as valid. More-senior workers may be reluctant to share knowledge if they feel that doing so undermines their status as experts.

In our telephone interviews of TD organizations, we found that some of these motivational factors were more predominant than others. Primary concerns were the level of effort required to codify content so that others could use it and free riding. Many were concerned that the effort required to share training content would not be commensurate with the benefit received. However, respondents were not concerned about losing credit for ideas, losing status as experts, becoming expendable, or being criticized by others.

Knowledge Use

Other factors can inhibit the use of knowledge that others have shared. The effort required for individuals to find information and effectively use it is a barrier to knowledge transfer. The concepts of "absorptive" and "retentive" capacity are relevant here.

Absorptive capacity is the ability of workers or organizations to exploit outside sources of knowledge, whereas *retentive capacity* is the ability of workers or organizations to "institutionalize the utilization" of the acquired knowledge (Szulanski, 1996). To effectively use knowledge that others have shared, recipients must possess the absorptive and retentive capacity to invest resources to acquire and effectively use knowledge (Tsai, 2001; Szulanski, 1996; Gupta and Govindarajan, 2000; Evangelou and Karacapilidis, 2005). Concerns about becoming expendable also can affect an individual's willingness to use existing knowledge. Another factor is perceived reliability of the knowledge. If individuals lack sufficient information about the quality of the knowledge or the reputation of the contributor, they may be reluctant to use the knowledge.

Technical factors also can affect knowledge sharing and use. Organizations must have tools (e.g., KMSs) that can be used to share and retrieve knowledge, and the features of these tools should support effective knowledge transfer. Barriers to using KMSs include an inappropriate human-computer interface, a lack of help tools, and inconsistencies in language or terms (Hackos and Redish, 1998; Virvou, 1999; Chandrasekaran, Josepheson, and Benjamins, 1999; Evangelou and Karacapilidis, 2005).

The predominant concern about knowledge sharing turned out to be the effort required to effectively use knowledge. Stakeholders from TD organizations were not concerned about the effort needed to find information, the possibility of becoming expendable, the reliability of the knowledge or reputation of the contributors, or the availability of efficient KMSs.

Participants in our study appeared not to be overly concerned about KM issues because they perceived the ROI from reuse to be low. Most of the stakeholders in our study, as well as the organizations they represent, did not consider the training material they produced to be valuable intellectual assets (e.g., because they felt it would be more difficult to reuse the material in a new context than to develop new material independently). Because they did not believe reuse was viable or, in many cases, that it would even occur, they naturally had little fear of losing credit for ideas, losing status as experts, becoming expendable, and the like. Furthermore, because any training material they might reuse tended to originate within their own larger organizations, they would likely have less difficulty finding the information (e.g., no metadata searches would be required) and would naturally have less concern about information reliability or reputation of the contributor.

However, in the event that a reuse strategy is perceived as a more valuable method for producing content in the future, traditional KM concerns will likely increase. In this case, valuable intellectual assets would more likely be at stake for many of the stakeholders, and they might perceive knowledge sharing as putting more at stake. Moreover, disincentives surrounding the use of knowledge would likely increase if large, open repositories containing content from a wide range of organizations come into being. In that case, it could be more difficult to search through these repositories to find material of interest (compared with searching within an individual's own organization) and to verify its quality.

Thus, we see a real potential for KM-like incentive issues to become more prevalent as the learning object economy and its enablers mature.

Incentive Mechanisms for Reuse Might Be Created by a Variety of Strategies

Creating incentives is a key technique for motivating behavior in organizations. Both sources and recipients of training content may require incentives to effectively participate in reuse activities. For individual workers, it may be important to measure and reward the extent to which they share content (Liebowitz, 2003; Evangelou and Karacapilidis, 2005). None of the TD organizations we interviewed were engaged in this practice.

Another potential incentive is reputation enhancement for employees who make important contributions to KMS or networks (Evangelou and Karacapilidis, 2005). At least one TD organization benefited from having the author's name attached to the output and, as a result, was more willing to share knowledge. Two other organizations provided employee incentives by showcasing reuse successes.[6]

Entire organizations may need incentives to collaborate in reuse activities with other organizations. Some organizations may fear that sharing training content will weaken their market advantage; however, Dyer and Hatch (2006) suggest that it is possible for an organization to achieve a competitive advantage even when sharing proprietary information in a knowledge-sharing network of competitors. Toyota cultivated a network of learning and knowledge between itself and its automotive suppliers, many of which were competitors. It found that greater knowledge sharing on its part resulted in a faster rate of learning within the suppliers' Toyota-related manufacturing operations.

---

[6] Highly motivating incentives might even help overcome the low ROI perception among stakeholders.

Cultivating a favorable organizational culture can also counteract some of the factors that inhibit knowledge sharing and use. Cultivating a sense of shared purpose and identity among employees to increase their feeling of belonging to a community can enhance sharing (Dyer and Nobeoka, 2000; Evangelou and Karacapilidis, 2005). In DoD organizations, we found some support for the notion that shared values (e.g., commitment to providing soldiers with more training opportunities) can promote reuse and obviate extrinsic incentives.

Creating formal roles with responsibility for sharing and using content is one way to facilitate a culture of reuse (e.g., Liebowitz, 2003). One TD organization noted that the organization had experienced success with reuse after creating technical support positions and providing training to aid the implementation of reuse. Another organization created internal boundary-spanner, or "synthesizer," roles to facilitate internal collaboration between IT staff who ran the software and instructional designers who made use of the software's capabilities.

For a network of organizations, a lead organization may need to set an example by heavily subsidizing content during the initial stages of network formation (Dyer and Nobeoka, 2000). For example, in the case of Toyota's network, it supplied proprietary information on its entire production process to the network and offered free assistance to its suppliers (Dyer and Nobeoka, 2000). In return, suppliers were required to offer knowledge of their operations to the network or risk losing Toyota as a customer. Toyota also provided an incentive for suppliers to participate. When a transfer of knowledge resulted in a productivity increase for a supplier, Toyota did not demand an immediate price decrease from the supplier. Instead, Toyota recognized that productivity increases would benefit Toyota in the long run and allowed suppliers to collect the short-term gains (Dyer and Nobeoka, 2000).

Features of technical systems, such as a well-designed user interface, help tools, a common language, and a searchable directory that identifies experts by topic (Liebowitz, 2003), may all enhance transfer of training content. Several organizations were engaged in efforts to simplify the process of implementing a reuse strategy (e.g., by purchasing an LCMS). However, efforts had not always been successful.

Several organizations cited the existence of mandates or financial pressure to promote a reuse strategy. In one case, the requirement to reuse was built into contracts with custom content developers. In two other cases, money was taken out of an organization's budget in anticipation of reuse, a move intended to change workers' behaviors. Respondents suggested that such pressures, when used in conjunction with other incentive mechanisms (e.g., providing new support), could have the intended effect.[7] However, it was also suggested that when such pressures are used in isolation, the strategy was not successful. We discuss this issue further, below.

DoD Mandate to Reuse May Require Additional Incentives to Be Effective

One mechanism for addressing incentive issues is the use of a mandate to require certain behavior within an organization. We consider here the high-level directive that requires reuse efforts within DoD (Instruction 1322.26 [DoD, 2006]).

This mandate will undoubtedly lead to a larger DoD reuse repository (the ADL-R) and provide increased opportunities for reuse, but we believe that it cannot fully address the incentive issue. Even as the ADL-R gets larger, current disincentives—such as the perception of low value, lack of general knowledge about how to design for reuse, and the ease of complying only with the letter but not the spirit of the mandate—may decrease the quality of repository content and increase the difficulty of finding truly reusable material. The result could be low use of many or most repository materials, as well as delays in implementing new business processes that favor reuse. Such a negative outcome could damage the perceived value of the emerging ADL-R at a time when positive perceptions may be critical to its success as a marketplace for

---

[7] In our case study (see Appendix A), financial pressure to reuse was applied in the form of reduced budgets. This strategy appeared to work in that workers saw reuse as a strategy for dealing with budget pressures. However, workers also seemed to value reuse because they saw the strategy as making sense and potentially promoting efficiencies.

reuse. Thus, supporting initiatives aimed at creating additional incentives may well be a requirement if the DoD instruction is to succeed.

Enforcement is another issue that clouds the potential success of DoD Instruction 1322.26. All training developers are required to submit information on content they have developed to the ADL-R, but the enforcement mechanisms that are to be used are unclear. An earlier mandate, DoD Directive 1322.20 (DoD, 1991), requires submission of content to DAVIS/DITIS but leaves enforcement to the military branches. In consequence, only a small portion of the "required" material has actually been submitted.[8]

Recommendation: Stimulate Additional Incentive Mechanisms for Participation in Reuse Strategies

We recommend that ADL explore additional incentive mechanisms for participation in reuse strategies and then encourage their use by increasing awareness among TD organizations and by building those mechanisms into contracts.

In the area of education and increasing awareness, ADL might supply TD organizations with information on how to foster positive organizational values among employees and how to develop recognition systems (monetary and otherwise) internal to their organization. ADL might also further support training on how to design for reuse.

ADL might pursue more buy-in for reuse from custom content developers with appropriate policy and contract changes. In the area of policy, ADL might allow developer identification and contact information to be used in metadata, so that highly reusable repository postings can serve as advertising and marketing tools for custom content developers. Such a policy might also help address concerns about content reliability. In the area of contracts, requiring maintenance for a fixed fee as part of the original course contract would give contractors more of a stake in designing content for reuse. Further, in an era when

---

[8] This discussion benefited from the input of G. A. Redding of the Institute for Defense Analyses, who was a contributor to both instructions, 1322.20 and 1322.26.

third-party organizations are likely to encounter technical challenges in attempting to reuse, contracts might be designed that require original custom content development companies to support future reusers up to a given level. Such measures could encourage the emergence of alternative business models among developers, to the extent that specialists in reuse emerge as a distinct type of organization.[9]

In general, knowledge about how to construct incentive mechanisms appropriate for TD organizations could be pursued in the case studies or pilot demonstrations suggested in the discussion on economic returns in Chapter Three. In other words, incentive issues among stakeholders could be explored along with more financial-related ROI issues at the organizational level.

---

[9] We feel that the market for reuse is not yet mature enough (i.e., there is not yet enough interorganizational reuse) to warrant investigation of formal methods of compensating original producers for reuse at this time.

CHAPTER FIVE
Implementation Issues

This chapter analyzes issues related to implementing a reuse strategy within a TD organization. Our analysis drew from our interviews with TD organizations and the literature on both change management (as it applies to e-Learning) and interorganizational collaboration.

Our findings indicate that organizations are still early in the process of implementing reuse strategies, despite their relatively long interest in such a strategy. We found that implementation obstacles were, as per our original hypothesis, more important than were technical and incentive issues in determining levels of reuse. Our overall conclusion in this area is that processes to support reuse require extensive development, beginning with the need for improvements in strategic planning for reuse. We recommend that ADL further evolve its role as a resource center and identify additional opportunities to support the implementation of reuse initiatives.

Implementing a Reuse Strategy for e-Learning Requires Significant Change Management

To move from a model focused on course production to a model based on reuse of learning objects, a TD organization may have to undertake extensive change-management efforts. Significant internal changes might be needed—e.g., with regard to the instructional design approach, business model, organizational culture, degree of collaboration with other organizations, use of technology, technical standards, procedures, and other processes. In general, the required changes are greater when an RLO-based (rather than a top-down or bottom-up) approach is pursued, because such an approach requires fundamental changes in how courseware is designed.

The move toward a learning object model requires changes in many areas:

- *Instructional design approach.* Successful reuse via large public repositories requires a fundamental change in how digital learning is developed. Essentially, the course-centered approach must be replaced with an approach centered on a learning object or a sharable content object (SCO), which is similar to an RLO. While SCOs can be large in scope, they are often associated with individual lessons or single learning objectives. Moreover, a learning object approach requires content to be separated from context, structure, presentation, and pedagogy.[1] Further, a successful design for reuse requires a way of dividing up the training material (referred to as *factorization* in Chapter Three) that is likely to be broadly applicable. Internal processes for designing training have to change as a result.
- *Business model.* A reuse strategy requires methods for determining reuse potential in each project and making decisions about when and when not to invest in reuse. These decisions will be based on the specific goals of the organization and the internal ROI that can be achieved, given demand and resources available.
- *Collaboration with other organizations.* At the RLO or asset level, a successful reuse strategy involving multiple organizations requires collaboration and often the formation of communities for either planned or ad hoc reuse (at RLO or asset level). For top-down reuse, there must be an investment in up-front collaboration for proper courseware design.
- *Employment of new technologies.* A successful reuse strategy requires investment in new tools, such as an LCSM and additional author-

---

[1] For a further explanation of how the separations might occur, see the Web site of the Reusable Learning Project, funded by the National Science Foundation (Reusable Learning Project, undated).

ing tools, as well as in the integration of these new technologies within work processes. An RLO approach, in particular, can benefit from the use of emerging authoring technologies.
- *Establishing new procedures and technical standards.* Implementing a reuse strategy requires the establishment of internal company standards and procedures beyond what SCORM suggests. For example, a procedure will be needed to match types of learning content with preferred types of media. A set of standards will have to be established on the grain size for instructional content—e.g., will each SCO be a course, a lesson, or a terminal learning objective? More generally, there will have to be a set of policies, procedures, and standards on exactly how to design training for reuse.
- *Alignment of incentives.* Measures are needed to ensure that all stakeholders have an incentive in their role. Incentives must also address the free-rider problem, especially for RLO reuse. It is important to set policies in this area that help establish a culture of reuse.
- *Legal and security issues.* Issues involving rights, compensation, security, and their effects on access to learning objects have to be addressed. Such issues include digital rights, copyrights, contracts, and access rules. Existing law is sufficient to address these issues in the context of e-Learning, but organizations need to decide on and establish their own rules within the broader legal context. The nonprofit organization Creative Commons[2] has sought both to democratize and to simplify the process of protecting the rights of developers and users of digital content. It offers a spectrum of options for granting use of digital content, from "full copyright" (all rights reserved) to "public domain" (no rights reserved).

Many of these changes tend to be greater when implementing an RLO-based reuse strategy rather than a bottom-up or top-down reuse approach. For example, the need to define grain size is unique to the RLO-based approach. Changes in the organizational business model, instructional design practices, and legal practices also tend to

---

[2] See the Creative Commons Web site (Creative Commons, undated).

be greater with an RLO-based approach. Further, successfully implementing an RLO-based reuse strategy requires a level of collaboration and new associations with outside training organizations that most TD organizations have not experienced in the past. Thus, we conclude that the biggest changes will need to occur to implement the RLO-based reuse approach.

TD Organizations Identified Many Obstacles Relating to Implementation of a Reuse Strategy

We asked the TD organizations to describe obstacles to increasing reuse within their organization. Figure 5.1 shows the described implementation obstacles aggregated to the organizational level of detail. Obstacles related to technology, incentives, and financial issues are covered in earlier chapters and thus are not reported here.

Figure 5.1
Implementation Obstacles Reported by TD Organizations

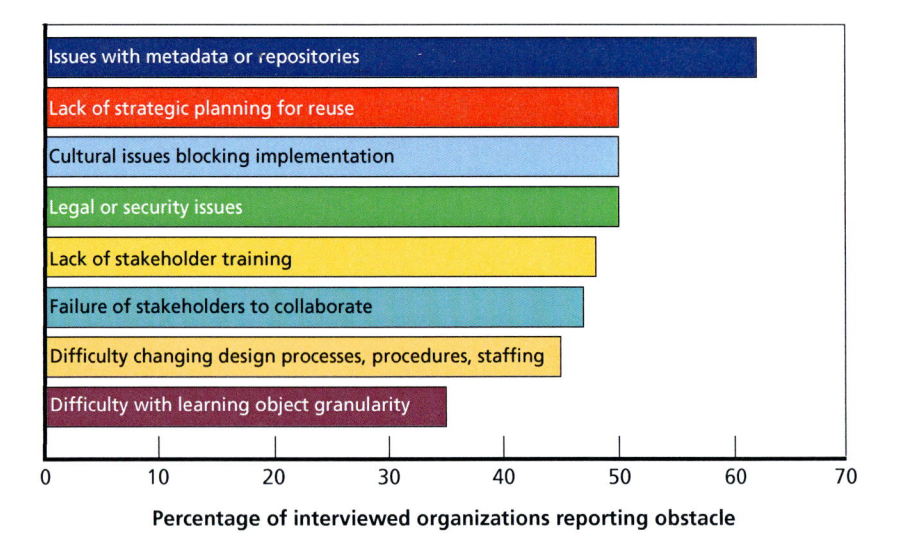

Obstacle	Percentage
Issues with metadata or repositories	~62
Lack of strategic planning for reuse	~55
Cultural issues blocking implementation	~52
Legal or security issues	~50
Lack of stakeholder training	~48
Failure of stakeholders to collaborate	~47
Difficulty changing design processes, procedures, staffing	~45
Difficulty with learning object granularity	~35

Percentage of interviewed organizations reporting obstacle

RAND MG732-5.1

The categories that the organizations identified are related to the organizational change process just described. As the figure indicates, TD organizations tend to experience a large number of obstacles; in fact, most of the bars in the graph show obstacles that were experienced 50 percent of the time. More specific detail on obstacles and comments by respondents are as follows.

Metadata or Repository-Related Issues
Nearly two-thirds of the organizations cited metadata or repositories as an obstacle or challenge to their strategy of reuse. As contributors to repositories, they noted the cost of establishing the metadata tags, especially when learning objects tended to be finely grained, such as at the level of the terminal learning objective. As potential users of the work of others, they noted that large public repositories had not yet proven to be effective vehicles for reuse and wondered whether the metadata tags that others attached to learning objects would be sufficient for finding the material they sought.

Lack of Strategic Planning for Reuse
Half of the organizations noted deficiencies in their larger organization's strategic planning for reuse. Strategic planning for reuse might be thought of as the disciplined effort required to formulate and make fundamental decisions about how to define and implement a policy of reuse. For example, a strategy would establish realistic and measurable goals for the amount of reuse that could be cost-effectively implemented. Strategic planning would also include formulating policies to support goal achievement and to establish and resource a comprehensive change-management process to enable success.

The strategic planning concept has been expressed as the "ends, ways, and means" paradigm. The "ways" are how an organization will employ its "means" (resources) to achieve its "ends" (goals). Respondents who cited strategic planning as an obstacle used each of the three terms in explaining their answers. Regarding the ends, respondents noted either a lack of support for or ignorance about reuse among the organization's leadership, or the absence of goals specified in a way that would allow success to be measured. Respondents also noted the lack of planning to determine how and in what cases reuse was to be achieved (e.g., how to change the up-front design process), and the absence of a set of much-needed procedures for implementing reuse. Finally, respondents noted the inadequacy of the resources provided to support the change process required for reuse.

Cultural Issues Blocking Implementation

Half of the organizations noted cultural issues that raised challenges for the adoption of reuse. Beyond what was described as a general resistance to change, the biggest cultural challenge was resistance to changing instructional system design practices in order to design for reuse. Another cultural block was a "not invented here" attitude toward material produced by others; in some organizational cultures, there is no tradition of sharing material. A third challenge was the tendency to see even material from nearly identical subject areas as "unique" and not a candidate for reuse; as one respondent stated, "Everyone wanted their own look and feel" for the training material. Finally, one respondent noted a skepticism among experienced military instructors about changes to training that would be implemented by "savvy young programmers" who could work the technology piece related to reuse but had little or no military experience.

Legal or Security Issues

Half of the organizations noted challenges with legal or security issues. Some reported struggling with copyright issues (e.g., outside the military, some vendors were unwilling or hesitant to hand over the rights to courses they had developed); others reported that permissions to use or reuse certain materials within an organization would not be adequate for placing material in a larger multiorganization repository. In addition, a number of military organizations reported difficulties in obtaining the valid declassifications of materials needed to make them generally available within the confines of a large repository. A few also reported problems in getting established permission systems to work properly, resulting in sign-in problems with DL courses.

Lack of Stakeholder Training

Nearly half of the organizations noted that their staff lacked training on how to implement reuse or that the organization lacked a plan for dealing with perceived shortcomings in staff skills. The biggest problem was instructional system designers' inability to effectively design for reuse. Respondents noted that designers were not being trained in how to use the authoring tools the organization purchased to enable reuse. Some respondents also noted a need for education on SCORM concepts and the potential for reuse, especially among the organization's leadership. This type of education was needed not only to sell the reuse concept within the organization, but also to avoid unrealistic expectations about what could be accomplished.

Failure of Stakeholders to Collaborate

Nearly half the organizations reported that the clear potential for reuse was not fully explored because the collaboration among stakeholders was poor. This was closely associated with both cultural issues (leaders resisted the idea of collaborating to obtain reuse) and problems with strategic planning (no process had established the need for collaboration). Some respondents noted that stakeholders were geographically fragmented or "in different stages" with regard to implementing SCORM, and others simply noted that stakeholders traditionally worked independently and liked the autonomy. Outside the military, several respondents noted that because their organization sought competitive advantage in the area of training, it did not explore what would otherwise be clear opportunities for collaboration.

Difficulty Changing Design Processes, Procedures, or Staffing

About 45 percent of organizations noted that although they had been willing to take all necessary action to accomplish reuse, implementation had fallen short because processes or procedures had not been established or required staff positions had not been filled. In several cases, organizations noted that implementation and testing of material designed for reuse added significantly to production time because adequate procedures for streamlining the process had not yet evolved.

Difficulty with Learning Object Granularity

Slightly more than one-third of the organizations mentioned the challenge of establishing the appropriate size of learning objects. A couple of organizations cited issues with finding the right balance between the increased reusability of smaller objects and the higher design cost of dividing learning into ever smaller chunks. Other organizations reported difficulty in implementing decisions about granularity. For example, a representative from one organization stated that the granularity defined in the organization's DL development contract was not always appropriate for learning needs. Another representative of an organization stated that the small grain size chosen for the SCOs led to complaints from trainees about the increased difficulty in finding the right learning segment.

What TD Organizations Saw as the Greatest Obstacle Depended on the Current Status of Their Reuse Efforts

To gain more perspective on the obstacles to reusing training content, we asked organizations to identify the single greatest obstacle among all the issues raised. We also categorized responses according to our assessment of whether an organization's reuse strategy was stalled or abandoned (which was the case in seven organizations) or was enjoying at least some successes (the case in 14 organizations). Figure 5.2 shows the distribution of greatest obstacles for both groups, including not only implementation issues, as discussed immediately above, but also issues dealing with technology, incentives, and ROI, which are discussed in more detail in previous chapters.

Figure 5.2 shows that implementation issues were most commonly cited as the greatest obstacle. For organizations whose reuse efforts were stalled or abandoned (pie chart on the left in the figure), the need for strategic planning and increased collaboration were the most significant obstacles (i.e., were noted by five of the seven organizations). In these cases, reuse was being implemented using the bottom-up approach, and the leadership either did not completely

Implementation Issues 73

Figure 5.2
Distribution of What Was Seen as the Greatest Obstacle to Reuse, by Status of Current Reuse Strategy

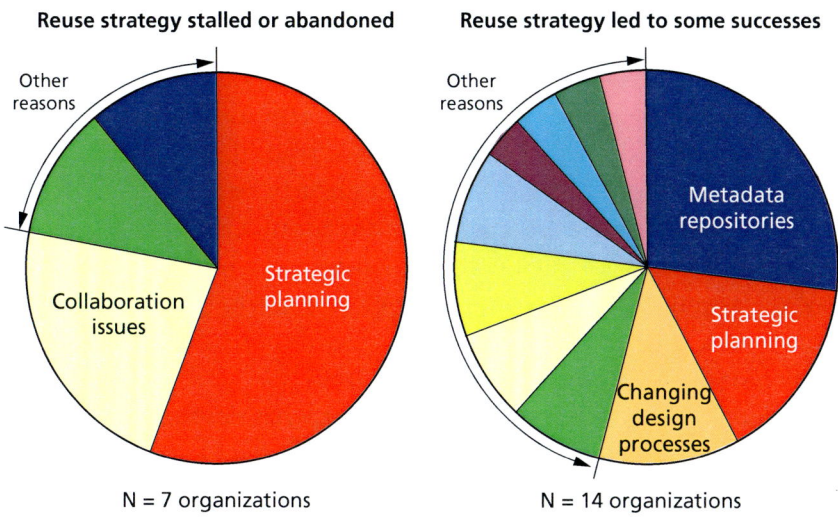

understand or did not firmly support reuse, or it supported reuse but either provided no planning resources to define where and how reuse would be accomplished or no resources to fund the up-front investment of the change-management process.

Strategic planning was frequently noted as the greatest challenge by organizations that had experienced some success (see pie chart on the right in Figure 5.2). In fact, representatives of one organization, which had invested heavily in reuse, felt that the organization's strategic planning had been deficient in many ways (e.g., they felt that the amount of possible sharing had been vastly overestimated).

For the 14 organizations that had experienced some success with reuse, however, the most notable obstacles were metadata and repositories. The lack of defined design processes was also an issue. These results suggest that organizations that had overcome some of the initial barriers to reuse were now looking to their next set of implementation challenges. Metadata and repositories made up a natural area in which to note challenges because they were the main focus of many organizations' current efforts at the time of the interviews.

The two pie charts in Figure 5.2 suggest that organizations are faced with a progression of challenges based on how far they have come in the change-management process. Some organizations were stuck at the beginning of a process (establishing a strategic plan), whereas others were focusing more on how to implement some of the details of a more established plan. Once large public repositories are established and working, organizations might well identify obstacles that are not occurring now because reuse is not by any means extensive. For example, issues to do with digital rights management or, as we suggested earlier, incentives may become more of a problem. Thus, a few years from now, a third pie chart may show a significantly different distribution of key obstacles for some organizations.

What appears clear from our study is that despite several years of effort, organizations are at a relatively early stage in the process of change, and much more development needs to occur.

A Supportive Environment Is Needed to Implement an Effective Reuse Strategy

Study participants' views are consistent with organizational experts' views emphasizing the need for a supportive environment in implementing an effective change-management process. (Lawler, 1992; Mohrman, Cohen, and Mohrman, 1995; Sundstrom et al., 1990). We discuss here three contexts for implementing a reuse strategy, one involving within-organization support and two involving across-organizations support.

Within TD organizations, support personnel are needed to facilitate collaboration between SMEs and technical staff. As discussed earlier and described in more detail in Appendix A, members of one organization emphasized the importance of creating internal "boundary-spanner" roles within their staff as a means to enable reuse. Other respondents noted the importance of finding technical staff who can collaborate successfully with educational experts on how to accomplish a goal. Such collaboration is successful when technical staff have some understanding of instructional systems design and, more generally, when individuals from different functional areas can "speak the same language."

A second important source of intra-organizational support is the availability of guidelines for how to design for reuse. These should cover various aspects of the implementation of an RLO-based reuse strategy—e.g., elements of a strategic plan and the business case for reuse within the organization, material that has a high likelihood of reuse in the TD organization, the proper granularity of learning objects for different situations, ways to reduce the up-front investment cost in realizing an RLO-based strategy, ways to re-engineer internal processes to foster reuse, and effective metadata schemes for an organization's training content.

Third, technical infrastructure is needed to make production cost-effective. Technical infrastructure consists of the hardware and software tools required to structure the development environment and deliver the training.

As described in Chapter Four, organizational culture is an important enabler for reuse. One component of successful organizational culture or cultural change is the presence of a champion for the effort. Staff are more likely to value reuse if a respected champion promotes its benefits than if senior managers merely mandate reuse. A champion also should ensure that the organizational context facilitates use by providing sufficient resources, effective training, and relevant policies, and by addressing the other facilitating factors already discussed in Chapters Two, Three, and Four. Additionally, the culture can be fostered by involving staff at all levels in the reuse plan. In particular, staff who are experienced and enthusiastic about reuse can provide technical support to others and serve as focal persons for collecting ideas and responding to issues that affect the organization.

As discussed in Chapter Four, incentives can be used to foster reuse and to facilitate cultural change more generally by rewarding valued behaviors. For instance, performance evaluations should include criteria associated with designing for reuse and reusing existing training content. Rewards, in the form of merit increases and/or bonuses, should be contingent on meeting these criteria. Examples of specific behaviors or outcomes that could be rewarded include

- contributing content to shared repositories
- contributing content to shared repositories that gets used by other staff and/or in other organizations
- creating content that meets standards that facilitate reuse
- reusing existing content
- contributing to guidelines or standards for reuse
- contributing to other organizational efforts that support reuse (e.g., training others to reuse or design for reuse, participating in strategic efforts for reuse)
- establishing internal or external social networks that lead to reuse.

Across TD organizations, the formation of communities of potential reusers can greatly increase the potential for reuse. Members of such a community need to work together to ensure that procedures and business rules regarding reuse can be established to meet as many as possible of the requirements of all members. For example, such communities might strive to formulate subject-specific standards and guidelines for reuse processes that can apply across organizations, along with a common language for metadata and metrics to measure success.[3] In addition, such communities may need rules or incentives to address the free-rider problem that can otherwise tend to reduce participation and reuse. Finally, such a community might enhance its success by trying to reach agreements that add infrastructure capable of supporting the collective development environment of all participating organizations. Such agreements, as implemented within a single organization, have already produced significant savings in the commercial sector. What

---

[3] For example, MedBiquitous states on its Web site (MedBiquitous, 2009) that it is a consortium striving to create "technology standards to advance healthcare education and connect the leading entities in professional medicine and healthcare." The goal of these standards is to better enable educators "to exchange educational content, track learner activities and profiles, and make healthcare education more accessible, measurable, and effective, thereby improving patient care."

can be accomplished in an interorganizational partnership remains to be demonstrated.

In some cases, different subject matter communities collaborate with each other to promote reuse. Such is the case with the technical documentation community and the training community within DoD. The training community uses SCORM standards, and the technical documentation community uses what is called "S1000D" standards. While technical publications and e-Learning have different goals and requirements, each largely begins with the same underlying content. The two communities are now collaborating to create standards that would allow the production of dual-purpose content, allowing use by multiple efficiencies within DoD. Large savings could occur, for example, in supporting training and documentation needs involved with the acquisition of new equipment (Katz, 2006).

Guidelines are also needed to help specific organizations that engage in top-down, strategic design for reuse—i.e., that co-produce courses. These organizations must establish partnerships, form design teams, create content that meets the needs of multiple audiences, and meet performance goals (e.g., timelines). The VA provides one example of such guidelines for e-Learning in the medical and military contexts.[4]

Widespread Success with Reuse Requires Effective Collaboration Across Organizations

Such cross-organizational activities as building repository communities and co-producing courses involve a number of steps and can entail significant transaction costs for participants. These activities involve finding and developing relationships with partners, negotiating contracts, establishing governance structures and other coordination processes, monitoring ongoing performance, and dealing with contract infractions or other unmet expectations. Clearly, these processes require substantial interorganizational communication and can generate conflict

---

[4] See discussion at the end of Chapter Three.

among partners and the need for negotiation. Larger numbers of participating organizations and significant differences in organizational cultures, priorities, and procedures will increase the potential for conflict, thereby increasing these costs.

There are currently few examples of interorganizational collaboration among training organizations. One exception is the VA's effort to develop reusable healthcare training content that has applicability across multiple agencies. As part of that effort, researchers are beginning to identify challenges and to formulate insights on how to overcome those challenges inherent to successful interagency collaboration (Twitchell et al., 2007).

Research on interorganizational collaboration in other contexts can also provide lessons for the training community. For instance, numerous studies of this type of collaboration have documented the importance of trust in creating and maintaining collaborative efforts (Faerman, McCaffrey, and Van Slyke, 2001; Huxham and Vangen, 2000; Kumar and van Dissel, 1996; Ring and Van de Ven, 1994; Snavely and Tracy, 2002; Vangen and Huxham, 2003). Trust emerges through interpersonal interactions, and establishing trust can take a substantial amount of effort and time. Research (e.g., Moss-Kanter, 1994) has identified a variety of organizational and interpersonal mechanisms to help achieve integration. Examples include

- ensuring contact among top leaders to develop strategic goals and objectives
- promoting interaction at mid-management levels to develop plans and identify resources
- identifying procedures and providing access to information and other resources for people at the operational level
- developing networks of interpersonal ties among participants
- providing participants with training in communication, conflict resolution, and cultural awareness
- providing participants with technologies to facilitate information exchange
- having individuals serve in boundary-spanning roles to facilitate communication

- identifying potential gaps to prevent or reduce conflict
- using a neutral third party for management and oversight of the collaboration.

Recommendation: Provide Additional Support for Processes That Implement Reuse Strategies

We recommend that ADL identify opportunities to support the implementation of reuse initiatives and further evolve its role as a resource center and consulting organization. Possible actions and roles for ADL include the following:

- Take the lead in facilitating the creation of new reuse communities. For instance, serve as a liaison to maintain a database of organizations' profiles and to broker relationships among organizations with strong potential to collaborate for reuse (e.g., because they serve similar populations of students).
- Create opportunities for people and organizations to form relationships/social networks by sponsoring conferences, workshops, mailing list servers, and other vehicles for establishing and maintaining communities of practice.
- Provide consulting services to organizations that are collaborating to reuse training content. Consultants could assist such organizations with contracts, developing standards, training for reuse, communication, conflict resolution, and other activities to support reuse.
- Create automated processes to monitor use of shared resources (e.g., to assess patterns of contributions to repositories, database searches, and use of content to determine whether participants' use of shared content is commensurate with their contributions; to assess the level at which participants tag content).
- Provide resources, such as model contracts or design templates, for organizations seeking to collaborate.
- Identify models of ways to develop training to meet the needs of different audiences.

- Identify high-profile pilots to document lessons learned about the effective (and ineffective) implementation of reuse. Collect lessons learned through surveys, interviews, and/or a central repository. Consider making services (e.g., consulting, brokering relationships) contingent on participation in data-gathering activities related to lessons learned.
- Provide recognition and tangible rewards for successful reuse activities and outcomes.
- Develop process metrics to support continual improvements. By creating a common set of metrics, organizations can have standards for assessing the impact of reuse.

The U.S. Department of Agriculture's (USDA's) establishment and ongoing operation of its "extension" agriculture consulting services, the Cooperative State Research, Education, and Extension Service (CSREES), is an interesting precedent for the ADL role we are recommending.[5] This outreach began in 1915 as a way to "push" best practices into American farming.[6] To achieve the goal of targeted outreach, the USDA set up organizations housed at the land-grant colleges around the country to provide specific, tailored consulting to farmers for free. The USDA attributes large successes in the increased productivity of American agriculture over the years to the outreach of information and best practices (on, for example, hybrids, pest control, soil management, and technologies) via CSREES.

The close relationship that the agricultural extension service has with the geographically distributed universities that house the consulting organizations recognizes regional differences in best practices and allows consultants to be close to their customers. In ADL's case, the consulting capabilities could be housed with existing co-laboratories

---

[5] CSREES's mission is to advance knowledge for agriculture, the environment, human health and well-being, and communities by supporting research, education, and extension programs in the Land-Grant University System and other partner organizations. The most important tools for accomplishing this mission are various sources of funding and program leadership. See Cooperative State Research, Education, and Extension Service, 2009.

[6] The Smith-Lever Act of 1914 was the start of the Cooperative Extension services as part of the land-grant universities.

that specialize in training and education in different contexts—e.g., within the defense organization, within the commercial sector, and within academia.

CHAPTER SIX
Overall Recommendations

ADL should aid the development of the learning object economy by focusing on what we have determined is a key enabler—the perceived value of reuse. ADL can provide this service by helping to determine the true potential for reuse, by identifying conditions leading to reuse's greatest payoff, and by increasing support to early adopters of reuse. Our recommended approaches for ADL are to

- Broaden the definitions of reuse and redefine success using metrics and surveys.
- Invest in high-profile pilots to identify conditions with the highest potential payoffs for reuse.
- Conduct or sponsor research to evolve guidelines for implementing reuse strategies.
- Evolve ADL's role as a neutral trusted advisor to TD organizations by
  - developing guidelines to help organizations understand how to design reusable content
  - disseminating best practices
  - fostering specific communities of reuse
  - conducting outreach in the form of workshops, conferences, and direct consulting.

ADL might also sponsor research aimed at developing a better understanding of how reuse efforts can be supported. Possible projects in this area are as follows:

- *Evaluate approaches for improving search capabilities for digital training content.* Half the battle of increasing the reuse of training content is finding and providing timely access to that content in the first place. Creating and applying high-quality metadata tags is resource and time intensive, and many content developers do not believe that the current ROI justifies the level of effort required to fully support reuse. Although some aspects of the metadata tagging process may soon be automated, there are still constraints that prevent potential users from seeing/experiencing content to evaluate its appropriateness for the desired reuse. However, such powerful search engines as Google now offer capabilities for searching public domain source code that provide searchers with a number of benefits, including direct access to the content itself or, when that is not possible, to an index of the content.[1] ADL might evaluate different approaches to providing an improved search capability for digital training content and estimate the costs and benefits for different types of reuse (e.g., object reuse, asset reuse, concept reuse).
- *Develop additional metrics for ADL-R's scorecard to capture costs and benefits to both contributors and seekers of content.* ADL would like to document the benefits and costs of the ADL-R to the community of developers of Web-based training content. With such information, ADL could effectively evolve tools and search methods to provide greater value to content seekers and identify ways to minimize the burden of data collection on ADL-R users. ADL might begin by examining basic metrics used by other repositories/referatories (e.g., libraries, DAVIS/DITIS) in order to consider how these might be expanded to measure the goals, successes, and failures of ADL-R users. Metrics could also be developed to measure content submitters' reuse-related development goals and resource expenditures and the different types of content seekers' reuse goals and success/failure rates.
- *Evaluate the evolution of the DL supply chain over time and identify interventions to improve the process.* The evolution of the DL supply

---

[1] See Google, 2009.

chain needs to be evaluated over time to predict the interventions needed to speed up progress toward rapid production of high-quality content at low cost. The process of developing Web-based training content has evolved dramatically via improvements in IT infrastructure, which have contributed to reductions in the cost of training content, increases in content quality, and improvements in the speed of development/revision. ADL might work with members of the ADL community and commercial DL sector to define the DL supply chain generically and to determine what its elements/enablers (e.g., tools, infrastructure) have been over time, starting circa 1990 and tracing up to the present. Different possible drivers of performance could be analyzed to assess how ADL might speed development of the most-critical enablers and/or spread best practices to reduce the time and cost of DL development while improving its quality.

- *Carry out focused case studies.* Case studies focused on current, high-profile efforts to maximize reuse of training content can be used to document emerging lessons learned and to identify "sweet spots" for different types of reuse. There are many possible conditions under which a reuse strategy might succeed. The TD community has consistently stressed the need for guidance on where these sweet spots are and on how to effectively invest resources for different types of reuse. "Natural" experiments in how to develop training content for reuse are currently occurring in DoD—e.g., the Joint Strike Fighter and the Future Combat System. By investing in studies of how such reuse is being implemented and identifying emerging lessons learned, ADL can capture and communicate best practices and ways to avoid common pitfalls.
- *Characterize the sweet spots for educational reuse.* It would be useful to produce guidelines and a decision tool that would help project/program leaders determine the likelihood of successful reuse. Previous research suggests that successful payoffs for investments in reuse will be difficult to predict. Factors influencing the success of reuse include aspects of the potential market (e.g., its size, type, and diversity), and the subject matter's complexity, specificity, and dynamic nature. ADL might consider supporting research

to identify relevant lessons about reuse from the software development arena and from efforts to produce reusable educational material that have been successful.

APPENDIX A
Case Study Results

This appendix presents the results for a case study involving one of the two organizations that participated not just in the broad set of interviews that were conducted by telephone, but also in site visits and more in-depth semi-structured interviews. We chose these two organizations for the site visits based on their reports of greater success with reuse of training content compared with other organizations in our sample. We considered them to be in a better position to provide lessons learned about reuse, especially about reuse success and enablers.

For the case reported here, three researchers spent 1.5 days conducting interviews on site. Fifteen staff and contractors participated in the interviews; they were instructional systems designers, SMEs/instructors, project managers, program directors, and technical support personnel for courseware. We conducted seven 60-minute interview sessions, predominantly with two or three participants at a time. For most of the interviews, three RAND researchers were present—one to conduct the interview and two to take notes.

The interview questions focused on RLO-based reuse, although we also discussed other reuse strategies. Questions addressed the perceived success of RLO reuse, factors that account for success, obstacles to reuse, and the role of incentives in creating reusable content or reusing existing content. After the interview notes were transcribed, participants' responses were coded into categories representing common themes.

The following sections summarize the results.

Success of RLO Reuse

Despite this organization's greater success with reuse relative to other organizations in our sample, the site-visit participants reported that RLO reuse was relatively uncommon within the organization. Participants in five of the seven interview sessions described the organization's success as modest or unclear, using such terms as *piecemeal*, *mixed*, *sporadic*, *can't tell*, and *don't know*. The RLO reuse that was occurring typically involved repurposing of content. For instance, staff had taken modules from existing courses and repurposed them into stand-alone continuous learning modules in order to reach larger audiences. Participants described greater success with reuse of entire courses rather than with RLOs. However, the organization generally did not design courses for reuse using a top-down strategy. Instead, reuse of courses generally came about in a more ad hoc manner in which members of a network of organizations determined that they had courses of interest to multiple populations of students. Subsequently, the focal organization sent its courses to other organizations or redeployed other organizations' courses on its own LMS.

Although respondents did not report using a "re-write" strategy for reuse, many remarked that reuse of other organizations' content almost always required "tweaking" to make it compatible with their own LMS.

Reuse Enablers

Interview participants were asked to identify factors accounting for the success they had experienced with reuse. Internal collaboration was the most frequently mentioned such factor. A strategy in this organization that was distinct from the strategies of the larger set of organizations we studied was creation of internal "boundary-spanner," or "synthesizer," roles. For example, some individuals facilitated internal collaboration among IT staff who ran the software on the server and instructional designers who made use of the software's capabilities. In addition, help-desk staff relayed issues raised by students to technical staff responsible for the LMS. More generally, internal collaboration helped members of the organization learn who knew what.

Boundary spanning between organizations, in the form of both informal and formal (strategic) social networks, was also mentioned as a factor accounting for success. These strategic partnerships facilitated identification and sharing of entire courses relevant to multiple organizations' student populations.

Thus, collaboration, both internal and external to the organization, was cited as a key driver of successful reuse. This finding is consistent what we found in the telephone interviews, which was that *lack* of collaboration was a significant factor in the stalling or abandonment of reuse strategies.

Most site-visit interview participants reported that lack of strategic planning was a key barrier to reuse (this is discussed in more detail below and in Chapter Five). This result is consistent with the results of the telephone interviews, in which strategic planning was identified as a key contributing factor to reuse success, and lack of strategic planning was identified as a key contributing factor to lack of success. Two site-visit participants mentioned up-front planning's contribution to reuse. One, for example, reported that when the organization had used other organizations' content, it had "gotten smarter" about testing earlier in the process.

Organizational culture was also cited as a factor contributing to reuse success. Some site-visit interview participants communicated a high level of commitment to reuse. An understanding of the value of reusing content seemed to be shared across roles in the organization, with the exception of the SMEs. In contrast, across the broader set of interviews, many participants described cultural values—such as resistance to changing instructional design practices and a "not-invented here" attitude—that inhibited reuse in their organizations.

A few site-visit interview participants mentioned other enablers for reuse, including explicit guidelines for development, documentation of best practices, flexibility to modify contracts to meet contractors' needs, and a matrix organizational structure that promotes interaction among individuals in different roles across the organization.

We also asked about the effects of specific factors, such as financial resources, on reuse. Responses indicated that reuse was influenced by both the scarcity and the availability of resources: A fixed budget or a lack of financial resources created motivation for reuse, but financial resources were needed to support an infrastructure for reuse (e.g., to create boundary-spanner positions).

Finally, we asked the on-site participants whether direct incentives served to motivate reuse. Responses to two questions on this issue suggest that explicit rewards or incentives were not a primary motivational factor. First, in responding to the open-ended question about factors that account for the success of reuse, no participants mentioned rewards. Second, when asked if there were explicit, positive incentives for reusing content or creating reusable content, all participants replied no. Instead, they cited informal rewards, such as "pats on the back," "atta boys," and recognition for performance in general, not for reuse per se. A few participants reported implicit incentives, including peer pressure and organizational culture. (Peer pressure might be more accurately classified as negative reinforcement rather than as an incentive, however.) Overall, it appears that incentives did not play a major role in the success of reusing training content in this organization or in staff's motivation to reuse content. The existence of an organizational culture that values reuse and collaboration seemed to make explicit rewards unnecessary; it also appeared to reduce the negative impact of disincentives or obstacles, as discussed below. At the same time, however, we do not know the extent to which explicit positive incentives would further motivate reuse.

Obstacles to Reuse

Participants in our on-site interviews mentioned many and varied obstacles to reuse, several of which served as disincentives to reuse.

Unlike many of the other organizations that participated in our study, this organization found that technical factors presented a substantial barrier to reuse. The predominant technical obstacle was the LCMS that the organization had adopted, which did not meet users' expectations and hindered reuse. Virtually all participants mentioned the LCMS as a significant obstacle to reuse.

Other technical barriers to reuse were mentioned, as well, although less frequently. One obstacle to reuse was limited "findability" of content, which might have been caused by such factors as an inability to conduct effective keyword searches in the organization, lack of content repositories, or lack of knowledge about where to look for such information outside the organization. A second obstacle was the frequent change in technologies and standards, such as moving from SCORM 1.2 to SCORM 2004. A third impediment, one that was raised by the majority of participants, was ambiguity or inconsistency in standards. Specific issues pertained to the lack of a common definition or understanding of the granularity of objects, differences in technical standards among the services, and differences in metadata schemes.

In addition to technical obstacles, the most common barrier to reuse pertained to organizational strategy. Issues cited included lack of an organizational strategy for reuse, absence of a reuse policy, lack of clarity in roles, failure to designate an individual or team as responsible for the reuse concept, insufficient up-front planning (including standards for how to make courses reusable), lack of business models for reuse, and calls for increasing IMI that may not be pedagogically necessary or appropriate. These findings about strategic issues are consistent with the telephone-interview findings in which strategic planning emerged as one of the primary factors associated with successful reuse, and the lack of such planning was the key driver of a stalled or abandoned reuse strategy.

A number of participants also identified financial obstacles to reuse. One was the time and effort required to reuse content, which included the need to analyze whether content could be reused, the need for new positions or a "middle layer" in the organization to serve as liaisons or boundary spanners, the need to rewrite content and/or make it run on the organization's LMS, and the need to compensate for the shortage of labor available to engage in these efforts. Another barrier was the lack of metrics for measuring ROI, which makes it impossible to know whether the efforts involved in reuse are worth the cost. In short, participants reported that it cost less to start from scratch than to identify and reuse existing content.

In contrast to the broader set of interviewees, a large number of the site-visit interviewees reported that SMEs presented a barrier to reuse, noting a variety of specific problems. Some said that SMEs often thought they understood how to design courses for reuse when they indeed did not. Some reported that SMEs did not share organizational values regarding reuse. For example, participants stated that SMEs did not understand (or care about) the impact of late courseware changes on the development process. Similarly, some SMEs were said to exhibit "turf issues" based on not wanting to use content developed by someone else. Others reported that SMEs were resistant to DL. These issues, coupled with power differentials in the organization (SMEs have more power), led to cultural clashes with respect to reusing training content and created barriers to internal collaboration.

A number of participants mentioned additional obstacles to implementing reuse. Several reported that legal issues or the use of proprietary content inhibited the efficiency of reuse. A few described a conflict between instructional design theory and reuse. For example, some instructional design theories call for continuity in course content such that lessons build on one another (Clark, 2004). In contrast, to create RLOs, each object must be independent and stand on its own. Similarly, Howard (2000) stresses the importance of providing context for course content, a goal that is at odds with the goal of making content appropriate for multiple audiences. Finally, two participants mentioned that inertia inhibited reuse in that some staff were reluctant to change their approach to content development.

Other Observations

Site-visit interview participants shared a number of other interesting observations or lessons learned about reuse:

- The effect of SCORM on reuse is unclear. One participant mentioned that the search for reusable content is independent of

SCORM. As evidence, this participant reported that reuse of content from residence training (which is not subject to SCORM requirements) was higher than reuse of DL content. (However, this difference could have resulted from the fact that there are far more residence courses available as a source of training content.)
- Some participants claimed that SCORM is not useful for courses that teach advanced skills or include dynamic content because the length of the development time means that material is out of date by the time a course is completed. Instead, they suggested using detailed tagging only for courses that cover basic, and relatively static, content.
- A corollary to reducing TD costs by reusing content is to figure out what students have already learned.
- A participant suggested that excessive reuse of the same content should be avoided so that students do not see the same examples or images repeated across courses. This would require a system that tracks where content has been used in other courses.

In summary, this case study of one organization allowed us to explore further the barriers to and enablers of reuse. Despite this organization's greater success with reuse compared with other organizations in the study, however, it had experienced only relatively modest levels of RLO reuse. Its most common kind of reuse was deployment of entire courses. Modest results can be attributed to some of the same obstacles to reuse reported by the other organizations, including lack of strategic planning, absence of metrics to demonstrate ROI, and anticipation of low ROI given the up-front costs of reuse. Participants in the site visits at this organization also revealed some unique challenges, including SMEs' negative attitudes and skills and technical factors (the LCMS). Although our study did not focus on technical issues, such as standards, these factors need to be addressed if reuse is to become widespread.

This organization's relative success can be attributed to some distinct processes. Key to reuse was the creation of internal boundary-spanning roles that served both to facilitate collaboration among staff in different functional areas and to help members of the organization understand who knew what. Informal and formal (strategic) social networks also contributed to the organization's success, in this case by facilitating identification and sharing of courses relevant to partner organizations' customers. Finally, the focal organization demonstrated commitment to reuse, which it showed, in part, by allocating financial resources for up-front costs (e.g., for boundary-spanner roles).

More important, the organizational culture supported reuse. The results suggest that shared values among most staff promoted reuse of content, obviated explicit rewards, and reduced the negative impact of barriers and challenges to reuse. Future research should investigate whether explicit incentives promote greater reuse and/or motivate less enthusiastic staff (SMEs, in this case) to be more cooperative. Nevertheless, other organizations may benefit from lessons learned at this site and from adopting similar processes to promote reuse of training content.

APPENDIX B
Interview Protocol and Questions

This appendix contains three documents that summarize the interview protocol and questions we used for collecting data from TD organizations. Once a candidate organization and point of contact were identified, an email was sent explaining the study and asking the organization to participate. We also sent two sets of interview questions. The first set (Part 1), which focused on basic facts and straightforward questions of an attitudinal nature, was designed to be completed prior to a telephone interview. The pre-interview questions helped expedite the phone conversation and keep it within a reasonable time period. The second set (Part 2), which consisted of open-ended questions, was designed to facilitate a semi-structured interview format in which the follow-up questions asked were based on initial responses. Informed consent and other information of potential interest to the interviewees was included in the initial email and Part 2 of the questionnaire.

General Letter:

My name is Dr. Michael Shanley from the RAND Corporation, a non-profit research organization dedicated to improving policy and decisionmaking. RAND is conducting a research project for the Advanced Distributed Learning (ADL) Initiative focused on the reuse of digital training content. The aim of this project is to identify policies that will speed the development of strategies for reuse.

The purpose of this email is to request the participation of the (organization) in an interview exploring implementation experiences with regard to reuse. Your organization has been selected because of its extensive experience in producing digital training content and in attempting to design for reuse of that content. The interview would be with you, and others you might nominate, and would cover the Department of Veteran Affairs experience with reuse, and what are seen as either enablers or obstacles to reuse in your organization.

The interviews will be conducted via telephone by RAND researchers and will last from 60–90 minutes. Prior to the interview, we will ask you to complete a brief survey and return it via e-mail. To properly represent your organization, we would like to talk with one or more persons knowledgeable about designing digital training content for reuse. We would prefer to talk to multiple participants simultaneously or, if necessary, we can accommodate more than one interview.

I have enclosed two documents for your review. The first is the brief survey that we request you return prior to the interview. Only one person needs to complete this survey. The second document consists of many of the questions we will ask in the interview. These questions do not require a written response, but we suggest that you, and anyone else from your organization who is participating in the interview, review them in advance so that we can move through the interview more quickly. This document also specifies the consent protocol for the study as well as the procedures for protecting information disclosed during the interview.

Your organization's participation in this interview is entirely voluntary. We realize that you have many demands on your time, but your input is critical in helping the ADL Initiative understand how to increase efficiency in training development and reuse.

If you agree to participate, we would like to set up a suitable time at your earliest convenience. I, or my assistant, will contact you within the next week to schedule an interview with you and other appropriate staff from your organization. Or, feel free to contact me at (email address) with your availability. If you would like to discuss the project beforehand, or have questions about who should participate in the session, please do not hesitate to contact me. Thank you.

Sincerely,

Michael Shanley
Policy Researcher and Principal Investigator, "Reuse" Project
RAND Corporation (www.rand.org)
1776 Main Street
Santa Monica, CA 90401
310-393-0411, x7795

RAND Project Survey for Advanced Distributed Learning Initiative

Reusable Digital Training Content: Obstacles and Enablers

Part 1: Questions to Be Addressed Prior to Telephone Interview

TO BE COMPLETED AND RETURNED TO MIKES@RAND.ORG PRIOR TO TELEPHONE INTERVIEW

PLEASE FEEL FREE TO CONTACT
DR. MICHAEL SHANLEY
IF YOU HAVE ANY QUESTIONS,
EITHER BY EMAIL AT MIKES@RAND.ORG OR
BY PHONE AT 310-393-0411, EXT 7795

September 2006

The **RAND Corporation**, a non-profit research organization dedicated to improving policy and decisionmaking

Pre-Interview Questions for Organizations Responsible for Reuse of Digital Training Content

Instructions

To expedite the upcoming telephone interview, please answer as many of the following questions as you can before the session. Any remaining questions can be addressed during the telephone interview, or beforehand via email or phone.

INSTRUCTIONS: Use your mouse to click the box or type in other information that corresponds to your answer for each of the following questions. All question answers can be edited if you want to change your answer.

Then please send the completed questionnaire back as a Word document via email.

Thank you very much for your participation.

Questions

- 1. What are the names and role(s) of those you expect to participate in the upcoming telephone interview? Please list each name and include a short description, including organizational name, if appropriate.

Participant	Role
_____	_____
_____	_____
_____	_____
_____	_____

- 2. Considering digital training content designed and developed internal to your organization or partnership:

    2a. What is the approximate number of personnel currently working half time or more on the production of digital training content? _____

    2b. What is the approximate number of hours of digital training content your organization produced last year? _____

- 3. Considering digital training content developed using external contractors:

    3a. List the number of contractor organizations currently completing work for your organization: _____

    3b. What is the approximate number of hours of content produced last year? _____

Note: For the following questions, "reuse" is defined as use of existing digital content to produce new content, or application of existing content to a new context or setting.

- 4. About what year did your organization first begin actively planning for reuse of digital training content?

    Year _____

- 5. About what year did your organization first begin producing digital training content with significant reuse?

    Year _____

- 6. About how much content (number of hours) has your organization produced that reuses existing content, or that has been applied to contexts or settings outside formal learning?

    Hours _____

- 7. Do you have any formal mechanisms (e.g., metrics) in place to measure reuse (e.g., amount, cost, ROI)? If so, please describe.

Metric	Description
_____	_____
_____	_____
_____	_____

- 8. How important do you consider SCORM, and its associated technologies (e.g., repositories, tools), as enablers in making reuse possible? (choose one response per line)

Standards and Technologies that might help your organization reuse digital training content at this point	Not a factor	Helpful	Helpful—but could provide more support	"Critical"
SCORM as a standard				
Public repositories of learning objects				
Content management systems within your organization				
Authoring tools that provide SCORM support				

- 9. As of this point in time, do you think your organization has saved money with your reuse strategy? (choose one)

    ____ No, did not intend to save money
    ____ No, but none expected at this early stage
    ____ No, expected savings did not materialize
    ____ Yes, but not as much as expected
    ____ Yes, as expected
    ____ Yes, more than initially expected
    ____ Other (Please describe)
    _____
    _____
    _____

- 10. What are your future plans with regard to implementing a reuse strategy—on a scale of 1–10 from abandon (1) to full speed ahead (10)? (choose one)

1.	2.	3.	4.	5.	6.	7.	8.	9.	10.
Abandon plans for reuse									Full speed ahead on reuse

RAND Project Survey for Advanced Distributed Learning Initiative

Reusable Digital Training Content: Obstacles and Enablers

Part 2: Questions to Be Addressed in Telephone Interview

PLEASE HAVE ALL PARTICIPANTS REVIEW THIS DOCUMENT PRIOR TO THE TELEPHONE INTERVIEW

PLEASE FEEL FREE TO CONTACT
DR. MICHAEL SHANLEY
IF YOU HAVE ANY QUESTIONS,
EITHER BY EMAIL AT MIKES@RAND.ORG OR
BY PHONE AT 310-393-0411, EXT 7795

September 2006

The **RAND** Corporation, a non-profit research organization dedicated to improving policy and decisionmaking

Informed Consent Information

RAND Consent Language

- RAND will use the information you provide for research purposes only, and will not disclose your identity or information that identifies you to anyone outside of the research project, except as required by law or with your permission.
- No one, except the RAND research team, will have access to the information you provide. RAND will only produce summary information from our collective set of interviews.
- We will destroy all information that identifies you after the study has concluded.
- You do not have to participate in the interview, and you can stop at any time for any reason.
- Your participation or nonparticipation will not be reported to anyone.
- You should feel free to decline to discuss any topic that we raise.
- **Do you have any questions about the study?**
- **Do you agree to participate in the interview?**

If you have any specific questions about this research, you may contact:

Michael Shanley, Ph.D.
Policy Researcher & Principal Investigator
RAND
P.O. Box 2138
Santa Mônica, CA 90407-2138
Telephone: 310-393-0411, x7795

Email: mikes@rand.org

Tora K. Bikson, Ph.D.
Human Subjects Protection Committee
RAND
P.O. Box 2138
Santa Monica, CA 90407-2138
Telephone: 310-393-0411
FAX: 310-393-4818
Email: Tora_Bikson@rand.org

RAND's Evaluation Team is led by Drs. Michael Shanley and Matthew Lewis.

RAND's Assessment of Reuse of Digital Training Content

- Overview: The ADL Initiative and the DoD are sponsoring the RAND Corporation (a non-profit research organization) in a study of the reuse of digital training content. The RAND team will be conducting interviews during the fall and winter of 2006–2007, exploring implementation processes involved in reuse. A particular focus will be the incentives of various stakeholders playing a role in developing training content. The ADL intends to use the results to formulate policies that encourage a higher level of reuse.

- Purpose of interview: As part of our evaluation, we are conducting interviews with members of organizations who are responsible for or knowledgeable about reuse of digital training or educational content. We are interested in your experiences and in what you have found to be enablers for and obstacles to reuse. We are also interested in your opinions with regard to how to overcome the obstacles.

- RAND participants: Drs. Michael Shanley and Matthew Lewis will be leading the interviews. A Research Assistant may also be present for the purpose of recording accurate notes.

Questions for Users of Digital Training Content

For purposes of the following questions, "reuse" is defined as the use of existing digital content to produce new content, or the application of existing content to a new context or setting.

A. Background
- 1. Briefly describe the organizations, or parts of your larger organization, that are exploring or pursuing a policy of reusing digital training or educational content.

- 2. How would you briefly describe the training or educational mission of your organization, or the larger organization of which your organization is a part?

B. Reuse—Current Status and Enablers
- 3. Briefly describe your organization's goals for reuse of digital training content. For example: What benefits do you see for your organization in reuse? What content would you aim to reuse, and what would you not reuse?

- 4. For what types of reuse do you see a potential in your environment? You might think of types of reuse in the following way:
    - Redeploy: Reuse content "as is," but in different contexts or for different groups (e.g., make a course available to a wider group, or use content not only for training, but for an on-line "help" system).
    - Rearrange: Reorder learning objects to form a new module (e.g., in a refresher course, move much of initial course to a backup section).
    - Repurpose: Update an existing module, or produce different "versions" of a learning module for different audiences.
    - Rewrite: Borrow assets from different learning objects to create new learning objects for a substantially new module or course.

- 5. Do you attempt to consider potential users outside your existing larger organization when designing content for reuse? If "yes," please explain.

- 6. What changes has your organization made (or is planning to make) in processes or in the use of technologies in order to enable reuse? For example:
    - Developed new strategies or business rules
    - Changed policies or processes
    - Instituted training programs

- o Made decisions on learning object granularity
- o Created new staff roles or organizational entities
- o Adopted standards
- o Purchased new technologies
- o Addressed digital rights issues

- 7. Considering all changes you have made or are planning, what do you consider to be the most important enabler of your reuse strategy?

C. Obstacles to Reuse

If you consider the following aspects of implementing a reuse strategy:
- ___ Strategic planning for reuse
- ___ Changes in internal design processes/procedures/staffing
- ___ Establishing appropriate granularity of learning objects
- ___ Employment of new technologies
- ___ Financial (i.e., funding levels or ROI)
- ___ Meta data schemes or repositories
- ___ Legal (i.e., relating to digital rights issues, such as copyrights, contracts)
- ___ Incentives among stakeholders
- ___ Employment of technical standards (e.g., SCORM)
- ___ Cultural
- ___ Other (list)

- 8. Which of the above are obstacles to increasing reuse within your organization? (list all that apply)

- 9. Overall, which of these factors do you consider to be the greatest obstacle to reuse in your context, and why?

D. Improvements needed

- 10. For the obstacles you noted above, what further improvement(s) could potentially eliminate them or lessen their effect?

- 11. What do you consider to be the most important improvement to implement, and why?

E. Incentive issues

- 12. Who would you say are the stakeholders in the reuse strategy; that is, those either making or affected by decisions about reuse within your organization?
- 13. Do any stakeholders show less than full enthusiasm/commitment in their participation?
- 14. For each case, what, in your opinion, explains their reluctance to fully participate?
- 15. Describe any efforts you are aware of to introduce positive *incentive structures* for different stakeholders to encourage greater reuse.

F. Conclusion

- 16. As of this point in time, do you think you have saved money with your reuse strategy?
- 17. Do you anticipate future savings in development costs from reuse?
    - If no, why not?
        - If yes, what percentage savings in development costs do you think is possible with a strategy of reuse (nearest 10 percent)? (Best case for comparison is cost of development given no reuse.)
- 18. Overall, what is the most important lesson you have learned about reusing training content?
- 19. Do you know of any organizations or partnerships (in government, academia, or the private sector) that have had substantial experience in pursuing a policy of reuse?

References

Advanced Distributed Learning (2008). "SCORM 2004, 3rd Edition," Web page, last revised December 22, 2008. As of February 11, 2009: http://www.adlnet.gov/scorm/

Alavi, M., and D. E. Leidner (2001). "Research Commentary: Technology-Mediated Learning—A Call for Greater Depth and Breadth of Research." *Information Systems Research*, 12, 2001, 1–10.

Ambient Insight (2006). *Ambient Insight's Snapshot of the 2006–2011 US eLearning Market. Highlights and Executive Overview of the New Report: The US Market for Self-Paced eLearning Products and Services: 2006–2011 Forecast and Analysis.* June 2006. As of January 13, 2009, downloads at: http://www.ambientinsight.com/Resources/Documents/AmbientInsight_2006_US_eLearning_Market_Snapshot.pdf

Barritt, C., and F. Lee Alderman (2004). *Creating a Reusable Learning Objects Strategy: Leveraging Information and Learning in a Knowledge Economy.* Pfeiffer, 2004.

BestWebTraining.com (2008). Web site, revised January 4, 2008. As of February 11, 2009: http://www.bestwebtraining.com

Boehm, Barry W., Hans Dieter Rombach, Victor R. Basili, and Marvin V. Zelkowitz (2005). *Foundations of Empirical Software Engineering: The Legacy of Victor R. Basili.* Springer, 2005.

Boehm, B., and B. Scherlis (1992). "Megaprogramming," in *Proc. DARPA Software Technology Conference.* Arlington, Va.: Meridian Corp, 1992.

Brandon Hall Research (2006a). *Custom Content Developers 2006: A KnowledgeBase of 130+ Companies That Provide Courseware Development Services.* Sunnyvale, Calif.: Brandon Hall Research, 2006.

——— (2006b). *LCMS Knowledge Base.* Sunnyvale, Calif.: Brandon Hall Research, 2006.

——— (2006c). *LMS Knowledge Base, 2005–6.* Sunnyvale, Calif.: Brandon Hall Research, 2006.

——— (2007). *Custom Content Developers 2007: A KnowledgeBase of 130+ Companies That Provide Courseware Development Services.* Sunnyvale, Calif.: Brandon Hall Research, 2007.

Brooks, Frederick P. (1979). *The Mythical Man-Month.* Addison-Wesley, December 1979.

Brynjolfsson, Erik (1993). "The Productivity Paradox of Information Technology: Review and Assessment." *Communications of the ACM*, December 1993.

Brynjolfsson, Erik, and Lorin M. Hitt (2003). "Computing Productivity: Firm-Level Evidence." MIT Sloan Working Paper No. 4210-01. June 2003. As of January 13, 2009, abstract available (full document available for download to subscribers) at:
http://ssrn.com/abstract=290325

Byers, Tom (undated). Web page with information by and about Byers, undated. As of February 11, 2009:
http://ecorner.stanford.edu/authorMaterialInfo.html?mid=1563

"Causes of ERP Failures" (undated). Web page, undated. As of February 11, 2009, accessible via "ERP" link at:
http://www.sysoptima.com/

Chandrasekaran, B., J. Josepheson, and V. Benjamins (1999). "Ontologies: What Are They? Why Do We Need Them?" *IEEE Intelligent Systems*, 14(1), 1999, 20–26.

Chapman, Bryan (2007). *Reusability 2.0: The Key to Publishing Learning.* White paper. Chapman Alliance, April 2007.

Clark, R. E. (2004). *Design Document for a Guided Experiential Learning Course.* Final report on Contract DAAD 19-99-D-0046-0004, from U.S. Army Training and Doctrine Command to Institute for Creative Technology and the Rossier School of Education. 2004.

Clark, Richard E., David F. Feldon, Jeroen J. G. van Merriënboer, Kenneth Yates, and Sean Erly (2006). *Cognitive Task Analysis.* October 14, 2006. As of January 13, 2009, downloads at:
http://cogtech.usc.edu/publications/clark_etal_cognitive_task_analysis_chapter.pdf

Concurrent Technologies Corporation (2003). *U.S. Navy Proves Case for Reusable and Interoperable Content Developed to the Sharable Content Object Reference Model (SCORM): Case Study of Systems-Level Oil Spill Prevention Training.* June 16, 2003 2003. As of January 13, 2009, downloads at:
http://www.instructtech.ctc.com/pdf/Spill%20Project%20SCORM%20Case%20Study.pdf

Cooperative State Research, Education, and Extension Service (2009). U.S. Department of Agriculture Web site, last modified February 11, 2009. As of February 11, 2009:
http://www.csrees.usda.gov/

Course Technology (undated). "CourseTechnology: Leading the Way in IT Publishing," Web page, undated. As of February 11, 2009:
http://academic.cengage.com/coursetechnology/?CFID=7305494&CFTOKEN=97645320

Creative Commons (undated). Web site, undated. As of February 11, 2009:
http://creativecommons.org/

Davis, Paul K., and Robert H. Anderson (2003). *Improving the Composability of Department of Defense Models and Simulations.* MG-101-OSD. Santa Monica, Calif.: RAND Corporation, 2003. As of January 15, 2009:
http://www.rand.org/pubs/monographs/MG101/

Department of Defense (1991). "Development and Management of Interactive Courseware (ICW) for Military Training." DoD Instruction 1322.20 (incorporating change 1, November 16, 1994). March 14, 1991.

Department of Defense (2006). "Development, Management, and Delivery of Distributed Learning." DoD Instruction 1322.26. June 16, 2006.

Digital Dream Designs (undated). "3D Models," Web page, undated. As of February 11, 2009:
http://www.digitaldreamdesigns.com/3Dmodels.htm

DoD—*See* Department of Defense

DoD Instruction 1322.20—*See* Department of Defense (1991)

DoD Instruction 1322.26—*See* Department of Defense (2006)

Dyer, J. H., and N. W. Hatch (2006). "Relation-Specific Capabilities and Barriers to Knowledge Transfers: Creating Advantage Through Network Relationships." *Strategic Management Journal,* 27, 2006, 701–719.

Dyer, J. H., and K. Nobeoka (2000). "Creating and Managing a High-Performance Knowledge-Sharing Network: The Toyota Case." *Strategic Management Journal,* 21(3), Special Issue: Strategic Networks, 2000, 345–367.

"E-Business Insight—ERP, CRM and Supply Chain Management" (2005). Home page, ©2005. As of February 11, 2009:
http://www.sysoptima.com/

Evangelou, C., and N. Karacapilidis (2005). "On the Interaction Between Humans and Knowledge Management Systems: A Framework of Knowledge Sharing Catalysts." *Knowledge Management Research and Practice,* 3, 2005, 253–261.

Faerman, Sue R., David P. McCaffrey, and David M. Van Slyke (2001). "Understanding Interorganizational Cooperation: Public-Private Collaboration in Regulating Financial Market Innovation." *Organization Science*, 12(3), May–June 2001, 372–388.

Glass, Robert L. (2001). "What's Wrong with Software Reuse?" As of January 13, 2009:
http://www.stickyminds.com/sitewide.asp?Function=edetail&ObjectType=COL&ObjectId=2731

Google (2009). "/*Code Search*/ LABS," Web page, ©2009. As of February 11, 2009:
http://www.google.com/codesearch

Google Enterprise (2009). Web page, ©2009. As of February 11, 2009:
http://google.com/enterprise/

Google SketchUp (2009). Web site, ©2009. As of February 11, 2009:
http://sketchup.google.com/

Gupta, A. K., and V. Govindarajan (2000). "Knowledge Flows Within Multinational Corporations." *Strategic Management Journal*, 21(4), 2000, 473–496.

Hackos, J., and J. Redish (1998). *User and Task Analysis for Interface Design*. New York: John Wiley & Sons, 1998.

Headquarters, TRADOC—*See* Headquarters, U.S. Army Training and Doctrine Command

Headquarters, U.S. Army Training and Doctrine Command (2003). *Training: Multimedia Courseware Development Guide*. TRADOC Pamphlet 350-70-2 (supersedes TRADOC Pamphlet 350-70-2 dated 10 July 2001). Fort Monroe, Va.: TRADOC, June 26, 2003.

Howard, R. W. (2000). "Generalization and Transfer: An Interrelation of Paradigms and a Taxonomy of Knowledge Extension Processes." *Review of General Psychology*, 4, 2000, 211–237.

Huxham, C., and S. Vangen (2000). "Leadership in the Shaping and Implementation of Collaboration Agendas: How Things Happen in a (Not Quite) Joined-Up World." *The Academy of Management Journal*, 43(6), 2000, 1159–1175.

Johnson, Larry (2002). "Elusive Vision: Challenges Impeding the Learning Object Economy." Unpublished draft documenting proceedings of a two-day conference on the learning object economy held September 2002 in San Francisco, Calif. Used by permission.

Katz, Heather A., Jo MacDonald, Stephen Worsham, and Paul Haslam (2006). *S1000D/SCORM Redundancy Analysis and Conversion Guidelines: Final Report*. Centreville, Va.: Intelligent Decision Systems, Inc., August 16, 2006. As of January 19, 2008, available for download at: http://www.jointadlcolab.org/research/s1000dscorm/index.aspx

Kiczales, G., J. Lamping, A. Mendhekar, C. Maeda, C. Videira Lopes, J. M. Loingtier, and J. Irwin (1997). "Aspect-Oriented Programming." *Proc. of ECOOP*, 1997.

Koenig, Michael E. D. (1993). *Business Process Redesign and the Productivity Paradox*. As of January 14, 2009: http://web.simmons.edu/~chen/nit/NIT%2793/93-193-koen.html

Kumar, K., and H. G. van Dissel (1996). "Sustainable Collaboration: Managing Conflict and Cooperation in Interorganizational Systems." *MIS Quarterly*, 20(3), 1996, 279–300.

Lawler, E. E., III (1992). *The Ultimate Advantage: Creating the High-Involvement Organization*. San Francisco, Calif.: Jossey-Bass, 1992.

Liebowitz, J. (2003). "Aggressively Pursuing Knowledge Management over 2 Years: A Case Study at a US Government Organization." *Knowledge Management Research and Practice*, 1, 2003, 69–76.

MedBiquitous (2009). Website, 2001–2009. As of February 12, 2009: http://www.medbiq.org

Mohrman, S. A., S. G. Cohen, and A. M. Mohrman, Jr. (1995). *Designing Team-Based Organizations: New Forms for Knowledge Work*. San Francisco, Calif.: Jossey-Bass, 1995.

Moore, Geoffrey (2002). *Crossing the Chasm: Marketing and Selling High-Tech Products to Mainstream Customers*. 2nd revised edition (originally published in 1991). New York: Harper Business Essentials, 2002.

Moss-Kanter, R. (1994). "Collaborative Advantage: The Art of Alliances." *Harvard Business Review*, July–August, 1994, 96–108.

MOWAG (2009). Web site, ©1997–2009. As of February 11, 2009: http://www.mowag.ch/

Multimedia Educational Resource for Learning and Online Teaching (2008). Web site, ©1997–2008. As of February 11, 2009: http://www.merlot.org/merlot/index.htm

Neven, F., and E. Duval (2002). "Reusable Learning Objects: A Survey of LOM-Based Repositories." *Proc. ACM Multimedia* (Juan Les Pins, France, December 1–6), 2002.

Newsome, Bruce, Matthew W. Lewis, and Thomas Held (2007). *Speaking with a Commonality Language: A Lexicon for System and Component Development*. TR-481-A. Santa Monica, Calif.: RAND Corporation, 2007. As of January 14, 2009:
http://www.rand.org/pubs/technical_reports/TR481/

NROC Commons (undated). Web page, undated. As of February 11, 2009:
http://www.nrocnetwork.org/nroc_commons

Numerial Algorithms Group (2008). "NAG Fortran Library," Web page, 2008. As of February 11, 2009:
www.nag.co.uk/numeric/fl/FLdescription.asp

OER Commons (2009). Web site, ©2009. As of February 11, 2009:
http://www.oercommons.org

Page-Jones, M. (1980). *The Practical Guide to Structured Systems Design*. Yourdon Press, 1980.

Pfleeger, Shari (1988). "Measuring Increased Productivity." *Office Systems Research Journal*, 6(2), Spring 1988.

Pfleeger, Shari, and Norris J. Cline (1985). "A Model of Office Automation Benefits." *Office Systems Research Journal*, 3(2), Spring 1985.

Reusable Learning Project (undated). Web site, undated. As of February 11, 2009:
http://www.reusablelearning.org/

Ring, P. S., and A. H. Van de Ven (1994). "Developmental Processes of Cooperative Interorganizational Relationships." *The Academy of Management Review*, 19(1), 1994, 90–118.

Roach, Steven S. (1991). "Services Under Siege—The Restructuring Imperative." *Harvard Business Review*, 69(5), 1991, 82–91.

Schmidt, Douglas C. (2006). "Why Software Reuse Has Failed and How to Make It Work for You." Earlier version appeared in *C++ Report*, January 1999. As of January 14, 2009:
http://www.cs.wustl.edu/~schmidt/reuse-lessons.html

Snavely, K., and M. B. Tracy (2002). "Development of Trust in Rural Nonprofit Collaborations." *Nonprofit and Voluntary Sector Quarterly*, 31(1), 2002, 62–83.

Succi, Giancarlo, and Francesco Baruchelli (1997). "The Cost of Standardizing Components." *StandardView*, 5(2), June 1997, 61–65.

Sundstrom, E., K. P. De Meuse, D. G. Ancona, and D. Futrell (1990). "Work Teams: Applications and Effectiveness." *American Psychologist*, 45(2), 1990, 120–133.

Supercourse: Epidemiology, the Internet and Global Health (2009). Web site, last update February 10, 2009. As of February 11, 2009:
http://www.pitt.edu/~super1/

Szulanski, G. (1996). "Exploring Internal Stickiness: Impediments to the Transfer of Best Practice Within the Firm." *Strategic Management Journal*, Special Issue: Knowledge and the Firm, 17, 1996, 27–43.

The 3D Studio (undated). Web site, undated. As of February 11, 2009: http://www.the3dstudio.com

TrainingTools.com (undated). Web site, undated. As of February 11, 2009: http://www.trainingtools.com/

Tsai, W. (2001). "Knowledge Transfer in Intra-Organizational Networks: Effects of Network Position and Absorptive Capacity on Business Unit Innovation and Performance." *Academy of Management Journal*, 44(5), 2001, 996–1004.

TurboSquid (2009). Web site, ©2009. As of February 11, 2009: http://www.turbosquid.com/

Twitchell, David, and Rebecca Bodrero (2006). "Interagency Collaboration Produces Sharable Training." Paper presented at the Interservice/Industry Training, Simulation, and Education Conference (I/ITSEC) for the 2006 annual conference.

Twitchell, David G., Rebecca Bodrero, Marc Good, and Kathryn Burk (2007). "Overcoming Challenges to Successful Interagency Collaboration." *Performance Improvement*, Special Issue, 46(3), March 2007.

Vangen, S., and C. Huxham (2003). "Nurturing Collaborative Relations: Building Trust in Interorganizational Collaboration." *Journal of Applied Behavioral Science*, 39(1), 2003, 5–31.

Varian, H. R. (1992). *Microeconomic Analysis*. 3rd ed. W. W. Norton & Company, 1992.

Virvou, M. (1999). "Automatic Reasoning and Help About Human Errors in Using an Operating System." *Interacting with Computers*, 11, 1999, 545–573.

Wiederhold, G., P. Wegner, and S. Ceri (1992). "Towards Megaprogramming." *Communications of the ACM*, 33(11), November 1992, 89–99.

Yourdon, E., and L. Constantine (1979). *Structured Design: Fundamentals of a Discipline of Computer Programming and Design*. Prentice Hall, 1979.